Water Tower of the Yellow River in a Changing Climate
Toward an integrated assessment

T0231291

Water Tower of the Yellow River in a Changing Climate

Toward an integrated assessment

DISSERTATION

Submitted in fulfillment of the requirements of
the Board for Doctorates of Delft University of Technology
and of the Academic Board of the UNESCO-IHE Institute for Water Education for the
Degree of DOCTOR
to be defended in public on
Monday, 15 September 2014 at 12:30 hours
in Delft, the Netherlands

by

Yurong HU

Master of Science in Hydrology and Water Resources,
UNESCO-IHE Institute for Water Education
Delft, the Netherlands
born in Henan Province, China

This dissertation has been approved by the supervisor:
Prof. dr. S. Uhlenbrook

Composition of Doctoral Committee:

Chairman	Rector Magnificus TU Delft
Vice-Chairman	Rector UNESCO-IHE
Prof. dr. S. Uhlenbrook	UNESCO-IHE / TU Delft, supervisor
Prof. dr. J. Xia	Wuhan University / Chinese Academy of Sciences, China
Prof. dr. B. Su	Twente University / ITC
Prof. dr. W.G.M. Bastiaanssen	TU Delft
Prof. dr. A. Mynett	UNESCO-IHE / TU Delft
Dr. S. Maskey	UNESCO-IHE
Prof. dr. H.H.G Savenije	TU Delft / UNESCO-IHE, reserve member

This research was conducted under the auspices of the Graduate School for Socio-Economic and Natural Sciences of the Environment (SENSE).

Published by:
CRC Press/Balkema
PO Box 11320, 2301 EH Leiden, The Netherlands
e-mail: Pub.NL@taylorandfrancis.com
www.crcpress.com – www.taylorandfrancis.com

ISBN 978-1-138-02714-5 (Taylor & Francis Group)

Summary

Climate change due to increasing greenhouse gas emissions is likely to alter the hydrological cycle resulting in large impacts on water resources worldwide. Mountain regions are important sources of freshwater for the entire globe, but their role in global water resources could be significantly altered by climate change. Mountains are expected to be more sensitive and vulnerable to global climate change than other land surface at the same latitude owing to the highly heterogeneous physiographic and climatic settings. Furthermore, there is also evidence from observational and modelling studies for an elevation-dependent warming within some mountain regions. With the increasing certainty of global climate change, it is important to understand how climate will change in the 21st century and how these changes will impact water resources in these mountain regions. Our understanding of climate change and the associated impacts on water availability in mountains is restricted due to inadequacies in observations and models. This is also the case in the Yellow River source region (YRSR). The YRSR is often referred to as the water tower of the Yellow River as it contributes about 35% of the total annual runoff of the entire Yellow River. Located in the northeast Tibetan Plateau, a "climate change hot-spot" and one of the most sensitive areas to greenhouse gas (GHG)-induced global warming, the potential impacts of climate change on water resources in this region could be significant with unknown consequences for water availability in the entire Yellow River basin. The YRSR is relatively undisturbed by anthropogenic influences such as abstractions and damming, which enables the characterization of largely natural, climate-driven changes.

A growing number of studies suggest that the YRSR is experiencing warming and streamflow reduction in recent decades, which has drawn increasing attention about the future climate changes and their impacts on water availability. While most previous studies focused on historical changes in the mean values of hydroclimatic conditions, future climate change impacts were less explored. Additionally, compared to assessing the impact of a change in average hydroclimatic condition, changes in extremes were solely missing in this region in spite of high relevance of such events on our society. This study attempts to fill these research gaps by investigating the spatial and temporal variability of both recent and future climate change impacts with specific focus on extremes. An integrated approach is applied consisting of (i) statistical analysis of historic data, (ii) downscaling of large-scale climate projections and (iii) hydrological modelling. This study contributes towards an improved understanding of spatial and temporal variability of climate change impacts in the YRSR through four major topics.

The first topic focuses on the assessment of recent climate change impacts in the YRSR. Historical trends in a number of temperature, rainfall and streamflow indices representing both mean values and extreme events are analyzed over the last 50 years. The linkages between hydrological and climatic variables are also explored to better understand the nature of recent observed changes in hydrological variables. Significant warming trends have been observed for the whole study region. This warming is mainly attributed to the increase in the minimum temperature as a result of the increase in magnitude and decrease in frequency of low temperature events. In contrast to the temperature indices, the trends in rainfall indices are less distinct. However, on a basin scale increasing trends are observed in winter and spring rainfall. Conversely, the frequency and contribution of moderately heavy rainfall events to total rainfall show a significant decreasing trend in summer. In general, the YRSR is characterized by an overall tendency towards decreasing water availability, which is shown by decreasing trends in a number of indices in the observed discharge at the outlet of basin over the period 1959–2008. The hydrological variables studied are closely related to precipitation in the wet season (June, July, August and September), indicating that the widespread decrease in wet season precipitation is expected to be associated with significant decrease in streamflow. To conclude, this study shows that over the past decades the YRSR has become warmer and experienced some seasonally varying changes in rainfall, which also supports an emerging global picture of warming and the prevailing positive trends in winter rainfall extremes over the mid-latitudinal land areas of the Northern Hemisphere. The decreasing precipitation, particularly in the wet season, along with increasing temperature can be associated with pronounced decrease in water resources, posing a significant challenge to downstream water uses.

In the second topic, three statistical downscaling methods are compared with regard to their ability to downscale summer (June–September) daily precipitation to a network of 14 stations over the Yellow River source region from the NCEP/NCAR reanalysis data with the aim of constructing high-resolution regional precipitation scenarios for impact studies. The methods used are the Statistical Downscaling Model (SDSM), the Generalized LInear Model for daily CLIMate (GLIMCLIM) and the non-homogeneous Hidden Markov Model (NHMM). The methods are compared using several criteria, such as spatial dependence, wet and dry spell length distributions and inter-annual variability. In comparison with other two models, NHMM shows better performance in reproducing the spatial correlation structure, inter-annual variability and magnitude of the observed precipitation. But its performance is less satisfactory in reproducing observed wet and dry spell length distributions at some stations. SDSM and GLIMCLIM showed better performance in reproducing the temporal dependence than NHMM. These models are also applied to derive future scenarios for six precipitation indices for the period 2046-2065 using the predictors from two global climate models (GCMs; CGCM3 and ECHAM5) under the IPCC SRES A2, A1B and B1scenarios. There is a strong consensus among two GCMs, three downscaling methods and three emission scenarios in the precipitation change signal. Under the future climate scenarios considered, all parts of the study region would experience increases in rainfall totals and extremes that are statistically significant at most stations. The magnitude of the projected changes is more intense for the SDSM than for other two models, which indicates that climate projection based on results from only one downscaling method should be interpreted with caution. The increase in the magnitude of rainfall totals and extremes is also accompanied by an increase in their inter-annual variability.

In the third topic, we investigate possible changes in mean and extreme temperature indices and their elevation dependency over the YRSR for the two future periods 2046–2065 and 2081–2100 using statistically downscaled outputs from two CGMs under three IPCC SRES emission scenarios (A2, A1B and B1). The projections show that by the middle and end of the 21st century all parts of the study region may experience increases in both mean and extreme temperature in all seasons, along with an increase in the frequency of hot days and warm nights and decrease in frost days. By the end of the 21st century, inter-annual variability increases in the frequency of hot days and warm nights in all seasons. The frost days show decreasing inter-annual variability in spring and increasing one in summer. Six out of eight temperature indices in autumn show significant increasing changes with elevation.

The fourth topic presents a modelling study on the spatial and temporal variability of the future climate-induced hydrologic changes in the YRSR. A fully distributed, physically based hydrologic model (WaSiM) was employed to simulate baseline (1961-1990) and future (2046–2065 and 2081–2100) hydrologic regimes based on climate change scenarios. The climate chance scenarios are statistically downscaled from two GCM outputs under three emissions scenarios (B1, A1B and A2). All climate change projections used here show year-round increases in both precipitation and temperature, which result in significant increases in streamflow and evaporation on both annual and seasonal basis. High flow is expected to increase considerably in most projections, whereas low flow is expected to increase slightly.

Snow storage is projected to considerably decrease while the peak flow is likely to occur later. We also observe a significant increase in soil moisture on annual basis owing to increased precipitation. Overall, the projected increases in all the hydro-climatic variables considered are greater for the mid of the century than for the end of the century. The magnitude of the projected changes varies across the subbasins, and is different under different emission scenarios and GCMs, indicating the uncertainty involved in the impact analysis. Inconsistency of observed streamflow trends with future projections indicates that the recently observed streamflow trends cannot be used as an illustration of plausible expected future changes in the YRSR. Such inconsistency calls for an urgent need for research aiming to reconcile the historical changes with future projections.

This study has covered a wide range of topics and a number of relevant issues of hydrology, climate change and downscaling in mountain areas. The applied multi-disciplinary approach has clearly added value and provided new insights (e.g. multisite downscaling in a mountainous catchment, climate-induced changes in extremes) and opened many new avenues for scientific research in the future to be explored including investigating the potential feedbacks between land cover change and climate change and reconciling the observed trends with future projections. In general, the knowledge generated in this study can be used as the basis of local scale adaptive water resources management in a changing climate.

Samenvatting[1]

Klimaatverandering als gevolg van de toenemende uitstoot van broeikasgassen zal waarschijnlijk de hydrologische kringloop veranderen, hetgeen wereldwijd grote gevolgen heeft voor de watervoorraden. Berggebieden zijn belangrijke bronnen van zoet water voor de gehele wereld, maar hun rol in de wereldwijde watervoorziening zou significant kunnen wijzigen als gevolg van klimaatverandering. Berggebieden zijn naar verwachting gevoeliger en kwetsbaarder voor wereldwijde klimaatverandering dan andere landoppervlakken op dezelfde breedtegraad als gevolg van de bijzonder heterogene fysiografische en klimatologische parameters. Bovendien bestaan er aanwijzigingen uit waarnemingen en modelstudies voor een hoogte afhankelijke opwarming binnen sommige berggebieden. Met de toenemende zekerheid van een wereldwijde klimaatverandering, is het belangrijk om te begrijpen hoe het klimaat zal veranderen in de 21^e eeuw en hoe de veranderingen de watervoorraden in deze berggebieden zullen beïnvloeden. Ons begrip van klimaatverandering en de daarmee gepaard gaande gevolgen voor de beschikbaarheid van water in de bergen is beperkt als gevolg van tekortkomingen in de waarnemingen en de modellen. Dit geldt ook voor het brongebied van de Gele Rivier (Yellow River Source Region, YRSR). De YRSR wordt vaak aangeduid als de watertoren van de Gele Rivier, want hij draagt ongeveer 35% bij aan de totale jaarlijkse afvoer van het gehele Gele Rivier. Gelegen in het noordoostelijk Tibetaans Plateau, een "climate change hot-spot" en in een van de meest gevoelige gebieden voor broeikasgassen (BKG)-geïnduceerde opwarming van de aarde, kunnen de mogelijke gevolgen van klimaatverandering op de watervoorraden in deze regio aanzienlijk zijn, met onbekende consequenties voor de beschikbaarheid van water in het gehele stroomgebied van de Gele Rivier. De YRSR is relatief verschoond gebleven van antropogene invloeden zoals onttrekkingen en dammen, hetgeen de karakterisering van grotendeels natuurlijke, klimaat gedreven veranderingen mogelijk maakt. Een groeiend aantal studies suggereert dat de YRSR de laatste decaden te maken heeft met opwarming en afvoerreductie, waardoor in toenemende mate de aandacht werd gericht op de toekomstige klimaatveranderingen en de gevolgen daarvan voor de beschikbaarheid van water. Terwijl de meeste voorgaande studies gericht waren op historische veranderingen in gemiddelde waarden van hydro-klimatologische parameters, werden de toekomstige gevolgen van klimaatverandering minder onderzocht. Verder is, vergeleken met de vaststelling van de gevolgen van een verandering van de gemiddelde waarden van hydro-klimatologische parameters, volledig voorbij gegaan aan de verandering van extremen in dit gebied, in weerwil van de hoge relevantie van dergelijke gebeurtenissen op onze samenleving. Deze studie tracht het ontbrekende onderzoek aan te vullen door het bestuderen van de ruimtelijke en temporele variabiliteit van zowel recente en toekomstige gevolgen van de klimaatverandering met specifieke aandacht voor extremen. Hierbij wordt een geïntegreerde benadering toegepast bestaande uit (i) statistische analyse van historische gegevens, (ii) schaalverkleining van grootschalige klimaatprognoses en (iii) hydrologische modellering. Voor een beter begrip van de ruimtelijke en temporele variabiliteit van de gevolgen van klimaatverandering in de YRSR is deze studie opgesplitst in vier grote thema's.

Het eerste thema richt zich op de beoordeling van recente gevolgen van klimaatverandering in de YRSR. Van een aantal indicatoren van temperatuur, neerslag en afvoer in de afgelopen 50 jaar zijn de historische trends van zowel de gemiddelde waarden als de extremen geanalyseerd. De verbanden tussen hydrologische en klimatologische variabelen zijn ook onderzocht om een beter inzicht te krijgen in de aard van de recente

---

[1] This summary is translated to Dutch by Mr. Pieter de Laat, Associate professor in UNESCO-IHE.

waargenomen veranderingen in de hydrologische variabelen. Hierbij zijn significante trends in de opwarming van het gehele onderzoeksgebied waargenomen. Deze opwarming wordt hoofdzakelijk toegeschreven aan de verhoging van de minimum temperatuur als gevolg van de toename in grootte en afname in de frequentie van lage temperatuur gebeurtenissen. In tegenstelling tot de temperatuurindicatoren zijn de trends in de neerslagindicatoren minder duidelijk. Echter, op stroomgebiedsschaal zijn stijgende trends waargenomen in de winter- en voorjaarsneerslag. Daar tegenover staat dat de frequentie en de bijdrage van matig zware regenval aan de totale neerslag een significante dalende trend laat zien in de zomer. Over het geheel wordt de YRSR gekenmerkt door een algemene tendens van afnemende beschikbaarheid van water, wat tot uiting komt in dalende trends in een aantal indicatoren van de waargenomen afvoer aan de uitlaat van het stroomgebied in de periode 1959-2008. De onderzochte hydrologische variabelen zijn nauw verwant aan de neerslag in het regenseizoen (juni, juli, augustus en september), wat aangeeft dat de wijdverspreide daling van de hoeveelheid neerslag in het natte seizoen geassocieerd kan worden met een significante afname van de gebiedsafvoer. Tenslotte toont dit onderzoek aan dat in de afgelopen decennia de YRSR warmer is geworden en blootgesteld was aan seizoensgerelateerde, enigszins wisselende veranderingen in de neerslag, wat tevens ondersteuning biedt aan een opkomend mondiaal beeld van opwarming en de overheersende positieve trend in extreme neerslag in de winter voor gebieden gelegen in de gematigde breedtegraad. De afnemende neerslag, vooral in het natte seizoen, samen met de stijgende temperatuur kan worden geassocieerd met een uitgesproken afname van de watervoorraden, wat een belangrijke uitdaging vormt voor het stroomafwaarts watergebruik.

In het tweede thema worden drie statistische schaalverkleiningsmethoden vergeleken op hun vermogen om de dagneerslagen van de zomer (juni-september) te reduceren tot een netwerk van 14 stations over het brongebied van de Gele Rivier, uitgaande van de opnieuw geanalyseerde NCEP/NCAR gegevens met als doel de bouw van hoge-resolutie regionale neerslagscenario's voor effectonderzoek. De gebruikte methoden zijn het Statistisch Downscaling Model (SDSM), het Generalized LInear Model for daily CLIMate (GLIMCLIM) en de Non-homogeneous Hidden Markov Model (NHMM). De methoden worden vergeleken met behulp van een aantal criteria, zoals de ruimtelijke afhankelijkheid, de distributie van de lengte van natte en droge perioden en variabiliteit van jaar tot jaar. In vergelijking met de andere twee modellen, geeft de NHMM methode betere prestaties bij het weergeven van de ruimtelijke correlatiestructuur, de jaarlijkse variabiliteit en de hoeveelheid waargenomen neerslag. Maar de resultaten in het reproduceren van de distributie van de lengte van natte en droge perioden van sommige stations zijn minder bevredigend. De SDSM en GLIMCLIM modellen lieten betere prestaties zien dan NHMM bij het reproduceren van de temporele afhankelijkheid. Deze modellen werden ook gebruikt om toekomstige scenario's af te leiden voor zes neerslagindicatoren voor de periode 2046-2065 waarbij gebruik werd gemaakt van de voorspellingen van twee globale klimaatmodellen (GCMs; CGCM3 en ECHAM5) onder de IPCC SRES A2, A1B en B1scenario's. Er is een sterke overeenkomst tussen de twee GCMs, drie schaalverkleiningsmethoden en drie emissiescenario's in het neerslagveranderingssignaal. Voor alle in beschouwing genomen toekomstige klimaatscenario's, zouden alle delen van het onderzoeksgebied stijgingen van neerslagtotalen te zien geven en een toename van extremen die statistisch significant zijn voor de meeste stations. De omvang van de voorspelde veranderingen is intenser voor de SDSM dan voor de andere twee modellen, wat aangeeft dat klimaatprognoses gebaseerd op resultaten van slechts één schaalverkleiningsmethode met voorzichtigheid geïnterpreteerd moeten worden. De toename van de totale hoeveelheid neerslag en de extreme regenval gaat verder gepaard met een verhoging van hun jaarlijkse variabiliteit.

In het derde thema onderzoeken we mogelijke veranderingen in gemiddelde en extreme temperatuurindicatoren en hun hoogte-afhankelijkheid over het gebied van de YRSR voor twee toekomstige perioden 2046-2065 en 2081-2100, waarbij gebruik wordt gemaakt van statistisch neergeschaalde uitvoer van twee CGMs onder drie IPCC-SRES emissiescenario's (A2, A1B en B1). De prognoses laten zien dat in het midden en het einde van de 21^e eeuw alle delen van het onderzoeksgebied in alle seizoenen een stijging van zowel gemiddelde als extreme temperaturen zullen ervaren, samen met een toename in de frequentie van hete dagen en warme nachten, en afname van dagen met vorst. Tegen het einde van de 21^e eeuw neemt de jaarlijkse variabiliteit in alle seizoenen toe met betrekking tot de frequentie van hete dagen en warme nachten. De dagen met vorst tonen een dalende jaarlijkse variabiliteit in het voorjaar en een toenemende in de zomer. Zes van de acht temperatuurindicatoren in de herfst laten een significante toenemende verandering met de hoogte zien.

Het vierde thema behandelt een modelstudie naar de ruimtelijke en temporele variabiliteit van hydrologische veranderingen in de YRSR veroorzaakt door veranderingen in het toekomstige klimaat. Een volledig fysisch gebaseerd ruimtelijk hydrologisch model (WaSIM) werd gebruikt voor het simuleren van de historische situatie (1961-1990) en toekomstige (2046-2065 en 2081-2100) hydrologische regimes, gebaseerd op scenario's voor klimaatverandering. De scenario's voor klimaatverandering zijn statistisch neergeschaald vanuit twee GCM modelsimulaties onder drie emissiescenario's (B1, A1B en A2). Alle hier gebruikte klimaatverandering prognoses laten het hele jaar door een toename zien van zowel de neerslag als de temperatuur, met als gevolg een aanzienlijke stijging van zowel de jaarlijkse als de seizoensgebonden afvoer en verdamping. Hoge afvoeren zullen naar verwachting in de meeste prognoses aanzienlijk toenemen, terwijl lage afvoeren naar verwachting weinig toenemen. De accumulatie van sneeuw neemt naar verwachting aanzienlijk af, terwijl de piekafvoer waarschijnlijk later zal optreden. We zien ook een aanzienlijke toename in het bodemvocht op jaarbasis als gevolg van meer neerslag. Over het algemeen geldt dat de verwachte toename van alle in beschouwing genomen hydro-klimatologische variabelen groter is voor het midden van de eeuw dan voor het einde van de eeuw. De omvang van de verwachte veranderingen varieert over de deelstroomgebieden, en is verschillend onder verschillende emissiescenario's en GCMs, hetgeen een aanwijzing is voor de onzekerheid van de uitkomst van de effectenanalyse. Inconsistentie van waargenomen afvoertrends met prognoses voor de toekomst geeft aan dat de recent waargenomen afvoertrends niet gebruikt kunnen worden als een illustratie van plausibele, te verwachten toekomstige veranderingen in de YRSR. Een dergelijke inconsistentie duidt op een dringende behoefte aan onderzoek om de historische veranderingen met toekomstige prognoses in overeenstemming te brengen

Deze studie heeft betrekking op een breed scala aan onderwerpen en een aantal relevante kwesties van hydrologie, klimaatverandering en schaalverkleining in berggebieden. De toegepaste multidisciplinaire aanpak heeft duidelijk toegevoegde waarde, gaf nieuwe inzichten (bv. multisite schaalverkleining in een bergachtig stroomgebied, door het klimaat veroorzaakte veranderingen in extremen) en opende vele nieuwe mogelijkheden voor toekomstig wetenschappelijk onderzoek, waaronder een studie naar de mogelijkheden van de feedback tussen veranderingen in bodembedekking en klimaatverandering en het in overeenstemming brengen van waargenomen trends met prognoses voor de toekomst. In het algemeen kan de kennis die met dit onderzoek werd gegenereerd gebruikt worden als basis van een op lokale schaal aangepast waterbeheer in een veranderend klimaat.

Acknowledgements

First of all I would like to acknowledge a number of agencies that jointly supported the work presented in this thesis: UNESCO-IHE Institute for Water Education, Rijkswaterstaat (the Ministry of Transport, Public Works and Water Management), The Netherlands and Yellow River Conservancy Commission (YRCC), China. The Program for Climate Model Diagnosis and Intercomparison (PCMDI) and the WCRP's Working Group on Coupled Modelling (WGCM) is acknowledged for making the WCRP CMIP3 multi-model data set available. Second, I would like to express my gratitude to a number of colleagues who provided data that supported my work: Dr. Hongli Zhao and Dr. Yangwen Jia of China Institute of Water Resources and Hydropower Research, Dr. Jin Shuangyan and Mr. Chunqing Wang of Hydrological Bureau, YRCC.

From a more personal point of view I would first of all like to thank Dr. Yangxiao Zhou for giving me the opportunity to do this PhD research. I still remember our discussion in Zhengzhou where you came for interviewing PhD candidates. Your continuous support and encouragement were very much appreciated throughout these years.

My sincere and deepest gratitude goes to my promoter Prof. Stefan Uhlenbrook for his willingness and acceptance to supervise my work. His helpful and friendly nature always makes me feel comfortable to ask for any support and guidance anytime. His wise guidance, critical and innovative insights and wealth of broad knowledge always kept me working in the right way.

I am extremely grateful to my supervisor Dr. Shreedhar Maskey. His critical thinking, valuable suggestions, constructive comments, fruitful discussions and enlightening guidance through the years have brought me to the point of successfully completing this thesis. Thank you very much for always encouraging me to go forward and purse excellence in every component of this study, for meticulously reading every paper at least three times, and for your straightforward solutions to complex problems.

My sincere thanks go also to Dr. Richard Chandler (Department of Statistical Science, University College London), Dr. Sergey Kirshner (Department of Statistics, Purdue University) and Dr. Joerg Schulla (Pacific Climate Impacts Consortium, Victoria, British Columbia, Canada) for their rapid and valuable support for solving some issues raised during the implementation of the GLIMCLIM and NHMM downscaling models and WaSiM hydrological model, respectively.

During this study, I worked at YRCC office in Zhengzhou and at UNESCO-IHE, Delft, Netherlands where I have received great cooperation and continuous support from my colleagues and friends at YRCC and UNESCO-IHE. I am thankful to all of them and wish them all the best. Special thanks to Mr. Pieter de Laat for translating propositions and summary into Dutch.

Especially, I am grateful to my husband Lushun Wang for his continuous support and encouragement during this PhD period. I would like to say sorry to my son Dinghan Wang and ask for his forgiveness for not spending much time with him and supervising his study. Last but not least I would like to thank my parents for always encouraging me to pursue things that I like. Mum and dad, this thesis is proof that your investment in education has paid off.

Contents

Figures and Tables

List of figures

List of tables

1. Introduction

1.1. Background

There is growing scientific evidence that global climate has changed, is changing and will continue to change (IPCC, 2013, and references therein). The latest IPCC Fifth Assessment Report (IPCC, AR5) has concluded that the global mean surface temperature has risen by 0.89°C from 1901 to 2012 and is likely to exceed 1.5°C or even 2°C (depending on future greenhouse gas emissions) relative to 1850-1900 by the end of the 21st century. Based upon energy and moisture budget constraints, precipitation is expected to increase in the global mean as surface temperature rises (Liu and Allan, 2013). Furthermore, precipitation disparities between wet and dry regions and between wet and dry seasons are expected to intensify in response to anthropogenic climate change (Biasutti, 2013). Both observations and model simulations suggest that wet regions and seasons will get wetter, and that dry regions and seasons will get drier (Biasutti, 2013; Liu and Allan, 2013; Chou et al., 2013; Polson et al, 2013), although there may be regional exceptions. The increased rainfall contrast between wet and dry regions and between wet and dry seasons will have serious implications for water resource management.

The global hydrological cycle is a key component of Earth's climate system (Wu et al., 2013). Global warming is expected to intensify the hydrological cycle resulting in strong impacts on water resources in many regions of the world. One major effect of global climate change is the potential changes in variability and hence extreme events (Marengo et al., 2010). Extreme events such as heatwaves, heavy rain or snow events, floods and droughts are of major concern for society as their impact on society is large. Besides, there is growing evidence that the nature, scale and frequency of extreme events are changing and will change further due to climate change (Kharin and Zwiers, 2005; Tebaldi et al. 2006; IPCC, 2012). In particular, heat waves and high temperatures are shown to increase significantly in frequency and severity in a large number of regions in the world (Clark et al 2006, Fischer and Schär 2010). In 2012, the IPCC Special Report on Managing the Risks of Extreme Events and Disasters to Advance Climate Change Adaptation (SREX) concluded that there was medium confidence that the length and/or number of heatwaves had increased since the middle of the 20th century and that it was very likely that the length, frequency, and/or intensity of these events would increase over most land areas by the end of the 21st century (IPCC, 2012). In its latest report, the IPCC pointed out that it is very likely (probability > 90%) that heat waves will occur with a higher frequency and duration, and it is virtually certain (probability > 99%) that there will be more frequent hot and fewer cold temperature extremes over most land areas (IPCC, 2013).

Climate change and the induced impacts are expected to vary regionally, even locally, in their intensity, duration and areal extent. For instance, there are indications that in particular, coastal, high-latitudinal, and mountainous regions belong to the most affected and vulnerable areas (IPCC, 2007). Furthermore, developing countries and countries in transition like China will be more vulnerable to climate changes due to their economic, climatic and geographic settings. Changes in climate and the induced hydrologic impacts are already being observed all over China, such as decreased precipitation over North China (worsening the water shortage in the north) and increased frequency of both severe floods and droughts in southern China (Yu et al., 2004; Wang and Zhou, 2005; Zhai et al., 2005; Zhang et al., 2006; Piao et al., 2010). Future climate changes are expected to continue to alter the temporal and spatial distribution of water resources over China (Sun et al., 2006; Gao et al., 2008; Liu et al., 2010; Piao et al., 2010).

With the ever-increasing certainty of global warming, sound studies of the assessment of climate change impacts are needed to facilitate the development of regional scale adaptation and mitigation strategies. Such a need holds additional importance for the mountain regions where the observed or projected warmings are generally greater than in low-elevation regions (Diaz and Bradley, 1997; Beniston et al., 1997; Rangwala et al., 2009; Liu et al., 2009; Qin et al., 2009). Mountain regions are likely to be particularly vulnerable because of their relatively high sensitivity to global climate change, large climatic variability over short distance and the vital role for local and downstream water related activities (Immerzeel et al., 2009). The Himalayas and the Tibetan Plateau, the source of major Asian rivers (e.g.Yangtze, Yellow, Mekong, Brahmaputra, Ganges, Indus), directly and indirectly supply water to the most populous region of the world with more than two billion people (Rangwala et al., 2012). The Tibetan Plateau has been identified as a "climate change hot-spot" and one of the most sensitive areas to greenhouse gas (GHG)-induced global warming (Giorgi, 2006) due to its earlier and larger warming trend in comparison to the Northern Hemisphere average and the same latitudinal zone in the same period (Liu and Chen 2000). The impact of climate change on this Asian water tower is likely to be significant (Immerzeel et al., 2010).

1.2. Contemporary research needs

Located in the northeast Tibetan Plateau, the Yellow River source region (YRSR) is geographically unique, possesses highly variable climate and topography, and plays a critical role for downstream water supply. A growing number of evidences suggest that this region is experiencing warming and decreased precipitation over the last 50 years (Xie et al., 2004; Zhao et al., 2007). Surface air temperature in the YRSR has increased by 0.76 °C at a rate of 0.18°C/decade from 1960 to 2001 (Zhao et al., 2007), which appears intense with respect to overall global warming. Annual precipitation exhibits a significant downward trend since 1990, especially summer precipitation. Impacts of a changing climate on the mountain hydrology are already evident, such as reductions in surface runoff, number of lakes, glaciers and frozen soil. Zheng et al. (2007) found that annual streamflow in the YRSR exhibited a statistically non-significant decreasing trend from 1956 to 2000, coinciding with the decreasing precipitation in the wet season (June to September). Wang et al. (2001) reported shrinking or disappearing of more than 2,000 small lakes out of 4,077 in the Madoi county— known as "the thousands lakes county". The lake water area decreased at a rate of 0.54% decade^{-1} from the 1970s to the 1980s, and 9.25% decade^{-1} from the 1980s to the 1990s. The permafrost is degrading considerably in response to the temperature changes. The lower limit of permafrost has risen by 50-80 m. The average maximum depth of frost penetration has decreased by 0.1-0.2 m (Jin et al., 2009). Degradation of permafrost has led to a lowering of ground water levels, shrinking lakes and wetlands, and noticeable change of grassland ecosystems alpine meadows to steppes. These changes are likely to result in a series of ecological and environmental problems in this region.

As the major source of water for the whole basin, a change in water resource in the YRSR not only affects water availability in this region but also in the middle and lower reaches of the river. However, our knowledge of how climate change will affect the availability of water in this region is rather limited owing to inadequacies in observations and models as well as unknown future climate change. To the best of our knowledge, literature on the impacts of future climate change in the YRSR is very limited; e.g. Xu et al. (2009) investigated the response of streamflow to climate change in the headwater catchment of the Yellow River basin with a focus on mean flow only at the outlet of this catchment. Regional extremes have recently received increasing attention worldwide given the vulnerability of our societies to such events. However, detailed assessments of how climate will change in the 21st

century, and how these changes will impact hydrological extremes, such as floods and droughts, are sorely missing in the YRSR. In addition, many existing studies focus on mean monthly or annual river flow, very few studies have considered the impact of climate change on other hydrological parameters such as evaporation, soil moisture, and groundwater (Calanca et al., 2006; Jasper et al., 2006; Rössler et al., 2012). This is particularly the case for the YRSR, where a detailed assessment of climate change impacts on the above hydrologic parameters is lacking to date. Furthermore, in large river basin like the YRSR with complex terrain and geology, the future hydrologic changes could be highly varied. Nevertheless, assessments of the spatial variability of future hydrologic response are not yet available. Therefore, there is a pressing need to assess the spatiotemporal variability of the future climate changes and their associated hydrological impacts in this mountainous catchment, including both mean state and extremes, in order to provide scientific support for taking appropriate adaptation and mitigation measures. The direct or indirect use of outputs from general circulation models (GCMs) to drive a hydrologic model may greatly enhance our insight in the potential impacts of global climate change. This research aims to enhance our knowledge base about the implications of global climate change on hydrological processes and water resources in subbasins of the YRSR with specific regard to the extremes.

This study focuses on the impact of climate change (changes in precipitation and temperature) alone on water fluxes and resources where anthropogenic influences such as land use/land cover and water consumption are not considered. Future conditions related to land use/land cover and water consumption may differ in addition to projected changes in climate. We argue, however, that changes in climate forcing represent likely the largest signal for the largely uninhabited, high altitude and relatively prestine basin considered here (80% of the basin covered by natural grassland).

1.3. Research objectives and approach

The main objective of this study is to investigate the future climate change and its potential impacts on hydrology and water resources in the YRSR, while considering the uncertainty arising from the choice of GCM and emission scenarios.
The following research questions are addressed:
1. What are historical trends and variability of hydro-climatic variables in the region? Are the trends in hydrologic variables explained by trends in climatic variables?
2. What is the appropriate method for downscaling large-scale atmospheric variables from GCMs outputs to a river basin scale in this region?
3. What are the credible future climate change scenarios for the YRSR?
4. Is climate response elevation-dependent in this mountain region?
5. How will the future climate change affect water balance dynamics and the discharge regimes of the YRSR, including both mean values and extremes? Will there be differences within sub-basins?
6. How do different sources of uncertainty contribute to the overall uncertainty in assessment of climate change impacts on water resources in the basin?

To meet the objective above and address the individual research question, an integrated approach to climate change impact assessment is developed by linking statistical trend analysis, statistical downscaling model, and hydrological model within a single framework. The methodological framework followed in this study is schematised in Figure 1.1. First, a statistical trend analysis (the Mann–Kendall (MK) test) is performed to detect historical trends in temperature, precipitation and streamflow in the second half of the 20th century (1961-2006). Trends in streamflow and their association with the climate trends are explored using partial correlation analysis. Second, three different statistical downscaling methods are compared and evaluated in order to select the appropriate rainfall downscaling method for the YRSR and illustrate the uncertainty in rainfall projection arising from the choice of downscaling methods. The methods used are the Statistical Downscaling Model (SDSM), the Generalized LInear Model for daily CLIMate (GLIMCLIM), and the non-homogeneous Hidden Markov Model (NHMM). The NHMM is then selected to develop future rainfall projections at multiple stations simultaneously. The Statistical DownScalingModel (SDSM) is applied to investigate possible changes in mean and extreme temperature indices and their elevation dependency over the YRSR for the two future periods 2046–2065 and 2081–2100. Finally, a fully distributed, physically based hydrologic model (WaSiM) was employed to simulate baseline (1961-1990) and future (2046–2065 and 2081–2100) hydrologic regimes based on climate change scenarios derived from statistically downscaling two global climate models (GCMs) under three emissions scenarios (B1, A1B and A2). The outputs from two GCMs under three emissions scenarios are used to explore the uncertainty linked to choice of GCMs and emissions scenarios. A brief description of the methods used in this study is presented below. For details, see the relevant chapters of this thesis.

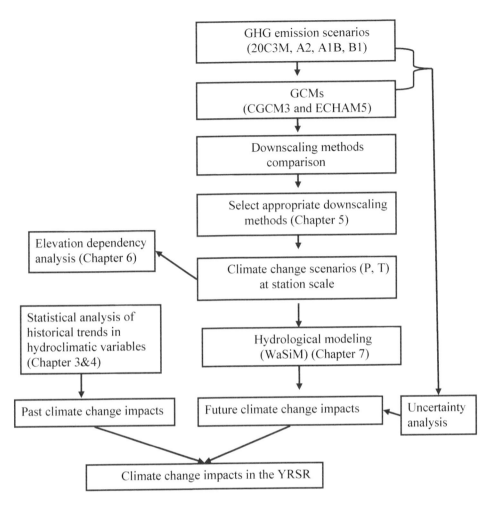

Figure 1.1: Methodological framework followed in this study.

1.4. Innovation and relevance

This thesis makes a contribution toward an improved understanding of changing hydrology in the YRSR and offers baseline information for adaptive water resources management in a changing climate. Specifically, this study presents a comprehensive modelling study on the spatial and temporal variability of climate change impacts in the YRSR. Quantification of potential climate change impacts in large mountainous catchment like the YRSR is particularly challenging due to the highly heterogeneous physiographic and climatological settings, and poor data availability. To our knowledge, this study is the first of its kind to address the question of the future hydro-climatic extremes and their spatial variability. With a comprehensive study on the hydrologic impacts of climate change for the YRSR, this research contributes to the scientific understanding of spatiotemporal variability of climate-induced hydrologic changes (both mean values and extremes). For a spatiotemporal evaluation of future hydrologic response, a multi-site downscaling model and a fully distributed, physically-based hydrological model are combined here for the first time. Such an approach is especially relevant for large mountainous catchment like the YRSR where physiographic and climatic characteristics vary considerably in space and time.

The results of this study could have important implications for water resources management in the basin. The knowledge generated by this study could serve as the basis of potential future directions of basin-wide adaptive water resources management and guide policy makers in taking appropriate, science-based action.

1.5. Thesis outline

This thesis research has resulted in four peer-reviewed international journal papers and one conference paper. The papers are, in modified form, included in this thesis as separate chapters. The thesis is structured as follows:

- Chapter 2 provides a brief description of the study area and the data sets used;
- Chapter 3 addresses recent historical trends in indices of rainfall and temperature extremes for the YRSR over the second half of twentieth century. This chapter is in its modified form published in Climatic Change (Hu et al., 2012);
- Chapter 4 investigates recent historical trends and variability in the hydrological regimes (both mean values and extreme events) and their links with the local climate in the YRSR over the last 50 years. Chapter 4 is in its modified form published in Hydrological Processes (Hu et al., 2011);
- Chapter 5 evaluates the performance of three statistical downscaling methods in reconstructing observed daily precipitation over the YRSR and presents future scenarios for six precipitation indices for the period 2046–2065 derived from statistically downscaling two GCMs (CGCM3 and ECHAM5) under three emissions scenarios (B1, A1B and A2). This chapter is in its modified form published in Theoretical and Applied Climatology (Hu et al., 2013).
- Chapter 6 investigates possible changes in mean and extreme temperature indices and their elevation dependency over the YRSR for the two future periods 2046–2065 and 2081–2100 based on the above mentioned two GCMs and three emission scenarios. Changes in interannual variability of mean and extreme temperature indices are also analyzed. Chapter 6 is in its modified form published in Hydrology and Earth System Science (Hu et al., 2013).
- Chapter 7 investigates the impacts of climate change on the hydrology of the YRSR in terms of streamflow, evaporation, soil moisture, and snow storage, as well as annual peak flow and 7-day low flow. Chapter 7 is in its modified form presented at the International Conference on "Climate Change, Water Resources and Disaster in Mountainous Regions: Building Resilience to Changing Climate" in Nov. 27-29, 2013, Kathmandu, Nepal. A related journal paper is submitted to Climatic Change.
- Chapter 8 forms the synthesis of the previous chapters, and presents a discussion of the main findings and further research needed in this area.

2. Study area and data sets

2.1. Basic hydroclimatology of the Yellow River source region

The YRSR is generally defined as the upstream catchment above the Tangnag hydrological station, situated between 95°50′45″E ~103°28′11″E and 32°12′1″~ 35°48′7″N in the northeast Qinghai-Tibetan Plateau (Figure 2.1). It covers an area of 121,972 km^2 (15% of the whole Yellow River basin), and yields an annual average runoff of 168 mm/a (35% of total runoff of the Yellow River). Therefore, it is called "water tower" of the Yellow River. It is characterized by highly variable topographic structure ranging from 6,282 m a.s.l. in the Anyemqen Mountains in the west to 2,546 m a.s.l. in the village of Tangnag in the east, which strongly influences the local climate variables and their spatial variability.

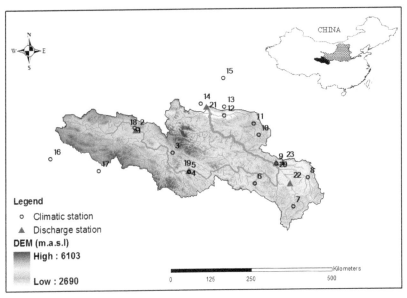

Figure 2.1: Digital elevation model of the study area showing the locations of hydroclimatic stations. Station numbers refer to Table 2.2. The smaller map in the upper right corner presents the location of the YRSR in China (black shaded area).

Climatically, the YRSR is cold, semi-humid characterized by the typical Qinghai-Tibetan Plateau climate system. The climate in this region is strongly governed by the Asian monsoon, which brings moist, warm air in the summer and dry, cool air during the winter (Lan et al., 2010). In winter, it has the characteristics of typical continental climate, which is controlled by the high pressure of the Qinghai-Tibetan Plateau lasting for about 7 months. During summer, it is affected by southwest monsoon, producing heat low pressure with abundant water vapour and a lot of rainfall and thus forms the Plateau sub-tropical humid monsoon climate. Annual average daily temperature varies between -4 °C and 2 °C from southeast to northeast. July is the warmest month, with a mean daily temperature of 8 °C. From October to April, the temperature remains well below 0 °C (Figure 2.2c). Mean annual precipitation ranges from 800 mm/a in the southeast to 200 mm/a in the northwest. Up to 75-90% of the total annual precipitation falls during the summer season (June to September) caused by the southwest monsoon from the Bay of Bengal in the Indian Ocean (Figure 2.2a).

In the months from November to March, more than 78 % of the total precipitation falls in the form of snow. However, the total amount of annual snowfall accounts for less than 10% of the annual precipitation. The rainfall in this region is generally of low intensity ($<$ 50mm/d), long duration (10-30 days) and covering large areas ($>$100,000 km^2). Mean annual potential evaporation varies from 800-1200 mm/a (Zheng et al., 2007). Similar to the precipitation, runoff in this region also undergoes large seasonal fluctuations consisting of a peak in July and a trough in February. Runoff from June to October accounts for 70% of the annual total (Figure 2.2b).

The spatial variability of soil and land use types is demonstrated in Figure 2.3. There are only grazing activities notable as human impacts. Grassland covers almost 80% of the region, and the total area of lakes and swamps is about 2000 km^2 (Zheng et al., 2009) (Figure 2.3, right). Eling and Zhaling, the two largest fresh water lakes in the region, cover 610 and 550 km^2, respectively. Snowpack and glaciers are present in the basin. The glacier coverage is about 0.16%, with the discharge contribution being less than 1% of the annual flow (Yang, 1991). Soils are mainly characterized by sandy loam and loamy texture (Figure 1.3, left).

Neither large dams nor large irrigation projects exist in this area, unlike the lower and middle Yellow River (Zheng et al., 2007; Sato et al., 2008). There is only one small size hydropower plant namely Huangheyuan located 17 km downstream of the Eling Lake with a maximum storage capacity of 15.2×10^8 m^3. The plant was constructed in 1998 and was put into operation in November 2001. However, the operation of the plant was halted between August 2003 and March 2005 because of insufficient inflow into the reservoir. The construction of the hydropower plant is expected to have some effects on the streamflow at Huangheyan and Jimai stations, while its effects on other downstream stations can be neglected as the annual mean flow at Huangheyan station only accounts for less than 5% of those of other downstream stations. Therefore, overall this study region is a relatively pristine area and has been subject to few human interventions.

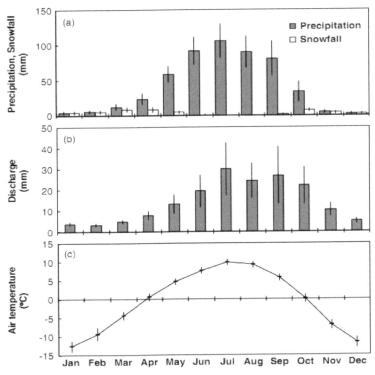

Figure 2.2: Monthly variations in (a) basin average precipitation and snowfall, (b) observed river discharge at the outlet, and (c) mean air temperature in the source area of the Yellow River from 1960 to 2000. (The error bars indicate the standard deviation for 41 years of data from 1960 to 2000. The areal values were calculated using the inverse distance-weighting method based on 16 climatic stations (adapted from Sato et al., 2008)).

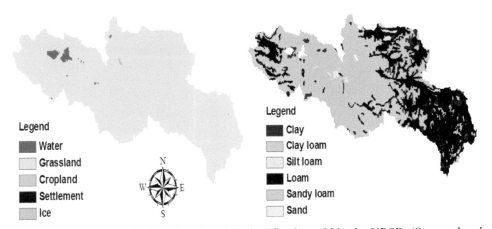

Figure 2.3: Land cover (left) and soil (right) classification within the YRSR. (Source: Land cover map: the Collection 5 MODIS Global Land Cover Type product; Soil map: the Harmonized World Soil Database (version 1.2)).

2.2. Overview of the data sets used in this study

Table 2.1 lists the major data sets used in this study and includes source information. Brief description of each of the data types is pesented in the following section. Further details are provided in the relevant chapters.

2.2.1. Climate and hydrology data

The study used daily observed hydroclimatic data for detection of recent trends, calibration/validation of the statistical downscaling models and the hydrologic model. Figure 2.1 shows the location of the 17 weather stations and 6 flow stations used in this study. Geographical characteristics of the hydroclimatic stations used in this study are displayed in Table 2.2.

2.2.2. Reanalysis data

For calibration and validation of the statistical downscaling models, large-scale atmospheric predictors are derived from the National Center for Environmental Prediction/National Centre for Atmospheric Research (NCEP/NCAR) reanalysis data set (Kalnay et al., 1996). This data set consists of specific humidity, air temperature, zonal and meridional wind speeds at various pressure levels and mean sea level pressure.

2.2.3. GCM data

In order to project future scenarios, outputs from two GCMs under the Intergovernmental Panel on Climate Change Special Report on Emissions Scenarios (IPCC-SRES) A2 (high-range emission), A1B (mid-range emission) and B1 (low-range emission) were used. These GCMs data are obtained from the Program for Climate Model Diagnosis and Intercomparison (PCMDI) website (http://www-pcmdi.llnl.gov). A detailed description of the GCMs is provided in section 5.2.1.

2.2.4. Spatial data

For the set up of the hydrologic model, the following spatial input data were used: (1) Digital Elevation Model as shown in Figure 2.1 based on the Shuttle Radar Topographic Mission (SRTM), version 4 (~90 m resolution; http://srtm.csi.cgiar.org/), (2) land use data as displayed in Figure 2.3 (left) based on the Collection 5 MODIS Global Land Cover Type product (~500 m resolution; http://lpdaac.usgs.gov/), and (3) soil data as displayed in Figure 2.3 (right)based on the Harmonized World Soil Data base (version 1.2) (~1 km resolution; http://webarchive.iiasa.ac.at/ Research/LUC/External-world-soil-database/HTML).

Table 2.1: An overview of main data sets used in this study

Category	Data set	Source
Climate	Precipitation, Temperature, relative humidity, relative sunshine duration, wind speed	China Meteorology Administration, Yellow River Conservancy Commission
Hydrology	Discharge	Yellow River Conservancy Commission
Reanalysis data	Large-scale atmospheric variables from the renanalysis	National Center for Environmental Prediction (NCEP)/National Centre for Atmospheric Research
GCM data	Large-scale atmospheric variables simulated from GCMs	Program for Climate Model Diagnosis and Intercomparison website
Topography	Digital elevation model (DEM)	Shuttle Radar Topographic Mission (version 4)
Land cover	Land cover map	Collection 5 MODIS Global Land Cover Type product
Soil	Digital map of the soils and soil properties	Harmonized World Soil Database (version 1.2)

Table 2.2: Geographical characteristics of the hydro-climatic stations used in this study

Station number	Station name	Latitude (North)	Longitude (East)	Elevation (m)	Data type
1	Huangheyan	34.95	98.13	4221	P
2	Maduo	34.92	98.22	4272	P, T_{max}, T_{min}, T_{mean}, WS, SH, RH
3	Renxiamu	34.27	99.20	4211	P, T_{max}, T_{min}, T_{mean}
4	Jimai	33.77	99.65	3969	P
5	Dari	33.75	99.65	3968	P, T_{max}, T_{min}, T_{mean}, WS, SH, RH
6	Jiuzhi	33.43	101.48	3628	P, T_{max}, T_{min}, T_{mean}, WS, SH, RH
7	Hongyuan	32.80	102.55	3491	P, T_{max}, T_{min}, T_{mean}, WS, SH, RH
8	Ruoergai	33.58	102.97	3439	P, T_{max}, T_{min}, T_{mean}, WS, SH, RH
9	Maqu	33.97	102.08	3400	P, T_{max}, T_{min}, T_{mean}, WS, SH, RH
10	Henan	34.73	101.60	3500	P, T_{max}, T_{min}, T_{mean}
11	Zeku	35.03	101.47	3663	P, T_{max}, T_{min}, T_{mean}
12	Tongde	35.27	100.65	3289	P, T_{max}, T_{min}, T_{mean}
13	Tangnag	35.50	100.65	2665	P
14	Xinghai	35.58	99.98	3245	P, T_{max}, T_{min}, T_{mean}, WS, SH, RH
15	Gonghe	36.27	100.62	2835	P
16	Qumalai	34.13	95.78	4231	T_{max}, T_{min}, T_{mean}
17	Qingshuihe	33.80	97.13	4418	T_{max}, T_{min}, T_{mean}
18	Huangheyan	34.88	98.17	4221	Q
19	Jimai	33.77	99.65	3969	Q
20	Maqu	33.97	102.08	3471	Q
21	Tangnag	35.5	100.15	2546	Q
22	Tangke	33.42	102.47	3470	Q
23	Dashui	33.98	102.27	3450	Q

P = precipitation [mm/d]; T_{max}, T_{mean} and T_{min} = daily maximum, mean and minimum temperature [°C]; WS = wind speed [m/s]; SH = sunshine duration [h]; RH = relative humidity [%]; Q = streamflow [m^3/s].

3. Trends in temperature and rainfall extremes in the YRSR[2]

Abstract: Spatial and temporal changes in daily temperature and rainfall indices are analyzed for the source region of Yellow River. Three periods are examined: 1960-1990, 1960-2000 and 1960–2006. Significant warming trends have been observed for the whole study region over all the three periods, particularly over the period 1960-2006. This warming is mainly attributed to a significant increase in the minimum temperature, which is the result of the significant increase in the magnitude and decrease in frequency of the low temperature events. In contrast to the temperature indices, no significant changes have been observed in the rainfall indices at the majority of stations. However, the rainfall shows noticeable increasing trends during winter and spring from a basin-wide point of view. Conversely, the frequency of moderately heavy rainfall events and contribution to the total rainfall in summer show a significant decreasing trend. To conclude, this study shows that over the past 40–45 years the source region of the Yellow River has become warmer and experienced some seasonally varying changes in rainfall, which also supports an emerging global picture of warming and the prevailing positive trends in winter rainfall extremes over the mid-latitudinal land areas of the Northern Hemisphere.

3.1. Introduction

In recent years, the frequent occurrence of extreme events, such as heat waves, heavy rain, hailstorm, snowfall, floods and droughts, have been reported worldwide (Ulbrich et al., 2003; Mirza, 2003; Kundzewicz et al., 2005; Kyselý 2008). This raised special concerns that the potential changes in the extreme events could accompany global climate change. There is great interest in assessing changes in extreme events because of their strong impacts on both human society and the natural environment. A number of theoretical modelling and empirical analyses have suggested that notable changes in the frequency and intensity of extreme events, including floods, may occur even when there are only small changes in the mean climate (Katz and Brown, 1992; Wagner, 1996). Groisman et al. (2005) showed that on a global scale, changes in heavy rainfall tend to be larger than changes in mean rainfall totals, and that increases in rainfall extremes occurred in many regions where no change or even a decrease in rainfall was observed. Changes in temperature and rainfall extremes in the twentieth century have also been observed in many parts of the world including the United States (Michaels et al., 2004), Canada (Vincent andMekis, 2006), UK (Osborn et al., 2000), Central and Western Europe (Moberg and Jones, 2005), Western Germany (Hundecha and Bardossy, 2005), Switzerland (Schmidli and Frei, 2005), Northeastern Iberian Peninsula (López-Moreno et al., 2009), Czech Republic (Kyselý, 2008), Italy (Brunetti et al. 2000, 2001), northeast Spain (Ramos and Martinez-Cassanovas, 2006), parts of Iran (Masih et al., 2010), China (Zhai et al., 1999; Zhai and Pan, 2003), India (Sen Roy and Balling, 2004), Mongolia (Nandintsetseg et al., 2007), Australia (Suppiah and Hennessy, 1998; Haylock and Nicholls, 2000), New Zealand (Salinger and Griffiths, 2001), South Africa (Kruger, 2006), and South East Asia and the South Pacific (Plummer et al., 1999; Manton et al., 2001). These studies, along with many others, are considered an important step towards knowledge of changes in climate extremes. Comparison between the results of various studies have demonstrated difference across region, with both increasing and decreasing or even no trends being reported. Possible reasons for lack of a clear picture of worldwide extreme events, e.g.

---

[2] This chapter is based on paper Trends in temperature and precipitation extremes in the Yellow River source region, China by Hu, Y., Maskey, S. and Uhlenbrook, S. 2012. Climatic Change 110: 403-429. DOI: 10.1007/s10584-011-0056-2.

regional climate variability, the different methods for trend-testing and the definition of the extreme events, are discussed in Wang et al. (2008). However, more importantly, these different findings suggest a large diversity in regional and global climate change interpretations, and imply that analyses of changes in extremes are important in the context of global climate change.

Over China, Zhai and Pan (2003) studied the changes in the frequency of extreme temperature events based on the daily surface air temperature data from about 200 stations from 1951–1999 in China. Their study showed a slightly decreasing trend in the number of hot days (over 35 °C) and the number of frost days (below 0 °C). Meanwhile, increasing trends were detected in the frequencies of warm days and warm nights, and decreasing trends were found in the frequencies of cool days and cool nights in China. Zhai et al. (2005) reported that rainfall indices showed mixed patterns of change, but significant increases in extreme rainfall have been found in western China, the mid-lower reaches of the Yangtze River, and parts of the southwestern and southern China coastal areas.

This study focuses on the source region of the Yellow River originating in the Tibetan Plateau. Located in mountainous areas, the study region is expected to be sensitive to global climate change since mountains in many parts of the world are very sensitive and susceptible to a changing climate in view of their complex orography and fragile ecosystem (Beniston et al., 1997; Beniston, 2003). Previous studies indicated that the Tibetan Plateau is one of the most sensitive areas in terms of response to global climate change (Liu and Chen, 2000; Tangetal., 2008) due to its earlier and larger warming trend in comparison to the Northern Hemisphere and the same latitudinal zone in the same period. Although many studies have been undertaken to investigate climatic changes in the Yellow River basin including its source region, most of them focused on changes in the mean values of climatic variables at monthly, seasonal and annual time scale. Assessments of extreme temperature and rainfall changes in the Yellow River reported in the literature are very limited. It is particularly true for the source region of the Yellow River. Furthermore, most of previous studies have concentrated on the second half of the twentieth century (1960–2000), and the recent years have been not included. For example, Fu et al. (2004) investigated the hydro-climatic trends of the Yellow River from the 1950s to 1998. They found that the river basin has become warmer, with a more significant increase in minimum temperature than in mean and maximum temperatures while the observed precipitation trend is not significant. Yang et al. (2004) analyzed the annual precipitation, mean temperature, pan evaporation, and river discharge trends in the Yellow River basin from the 1950s to the 1990s. It was found that the annual precipitation showed a non-significant decreasing trend of 45 mm and the air temperature increased by 1.28 °C over the past 50 years. Tang et al. (2008) analyzed the changes in the spatial patterns of climatic and vegetation condition in the Yellow River basin from 1960 to 2000 and found decreasing precipitation and increasing temperature in most parts of the Yellow River basin with the largest temperature increase in the Tibetan Plateau. Xu et al. (2007) investigated the long-term trends in major climatic variables in the Yellow River basin from1960-2000. The results indicated that temperature has increased over the Yellow River basin especially during autumn and winter, while the precipitation has generally decreased in all the seasons except in winter which shows a slight increase. They also found that precipitation in the upstream region of the Yellow River did not exhibit a significant trend, whereas both the middle and downstream regions showed a clear negative trend. This is also confirmed by Wang et al. (2001) who found that the annual precipitation in the headwater area showed no noticeable decreasing tendency between the 1950s and the 1990s. But summer precipitation (from June to September) showed a tendency to decline. Zhao et al. (2007) reported that over the past 40 years the annual mean temperature has significantly increased by 0.8 °C in the upper Yellow River Basin while annual precipitation slightly decreased by 43 mm. Zhang et al. (2008) recently published work on winter extreme low temperature events and summer extreme high-temperature events in the whole Yellow River for the period 1960–2004. They found that the whole Yellow River basin is dominated by the significant downward trend of frequency of the cold events and that significant upward trend of frequency and intensity of the high temperature events has been found in the western and northern part of the Yellow River basin.

As discussed earlier, most previous studies regarding climatic trends were based on the mean values of climatic variables. Although Zhang et al. (2008) studied the temperature extremes of the Yellow River for 1960-2004, they did not consider changes in rainfall extremes and in other important seasons such as spring (March-May) and autumn (September–November) in their study. The need for involving spring and autumn into studies of changes in climatic variables was highlighted by Kyselý (2008) since the differences in climatic variables exists not only between winter and summer but also between the transition seasons (spring and autumn).

In this study, we use daily temperature and rainfall data for the period 1960–2006 to study the spatial and temporal changes in various indices for rainfall and temperature extremes in the source region of the Yellow River. Our study complements previous work by including a longer time series of data and more climatic extreme variables in the analysis. Furthermore, we include annual as well as four seasonal analyses, whereas Zhang et al. (2008) only considered winter and summer. Therefore, to the best of our knowledge this paper is the first most comprehensive regional analysis of trends in indices of rainfall and temperature extremes for the YRSR over the second half of twentieth century based on historical climate observations. Although the main aim of the present study is to investigate trends in temperature and rainfall extremes in the Yellow River source region on both annual and seasonal basis, changes in the average climatic indexes are also assessed.

3.2. Data and methods

3.2.1. Data base

Daily rainfall totals measured at 14 stations and daily maximum and minimum temperature at 13 stations (Table 2.1, Figure 3.1, 3.6) operated mostly by Yellow River Conservancy Commission and China Meteorological Administration are used. The observations span the period 1960–2006 and 1961–2006 for rainfall and temperature, respectively. The available series lengths range from 30 to 47 years. In order to seek a more spatial coverage in this data sparse mountainous region and make comparisons between different periods, data were analyzed for three different periods. The duration of the periods analyzed were 31 years, 41 years and 47 years for rainfall as well as 30 years, 40 years and 46 years for temperature with each period starting in 1960 for rainfall and 1961 for temperature.

3.2.2. Data quality control

The objective of data quality control was to identify questionable records in the climate data sets. Several types of quality controls were applied to the series of daily rainfall, daily maximum temperature (Tmax) and daily minimum temperature (Tmin). First, the daily time series from each station were plotted and compared with neighboring stations for identifying outliers and missing data. Second, a revision of internal consistency was made, verifying that daily Tmax always exceeds daily Tmin. Third, annual mean series (annual total for rainfall) were produced from the daily rainfall and maximum and minimum temperature time-series, and examined for homogeneities using the double mass curve method, particularly to check if there are clear indications of relocation of the stations and/or change of instrumentation or observational practices.

Data gaps found in the time series were very minimal: 0.029% and 0.0017% of daily records for rainfall and temperature, respectively. The missing data values were filled up by their neighboring stations with the simple linear regressions, and the outliers were corrected with the data from the nearest stations or from the neighboring days. Results of the double mass curves of all stations demonstrated almost a straight line, thus no obvious breakpoints were detected in the time series of temperature and precipitation.

3.2.3. Indices for characterizing temperature and rainfall extremes

The indices chosen to evaluate changes in the rainfall and temperature patterns are able to represent a wide variety of rainfall and temperature characteristics for both the average regime and the extreme behavior of the rainfall and temperature processes. We selected 15 indices describing different aspects of the rainfall and temperature regimes. Many of them have been used in previous studies (Moberg and Jones, 2005; López-Moreno et al., 2009) and recommended by the STARDEX project (Haylock and Goodess, 2004). Table 3.1 provides the acronyms and short definitions of the selected indices. Values of each index were calculated on an annual basis and for four seasons: December to February (DJF), March to May (MAM), June to August (JJA) and September to November (SON). The STARDEX extremes indices software is available at http://www.cru.uea.ac.uk/cru/projects/stardex/ and was used for calculating all these indices in this study.

Table 3.1: Investigated indices of daily precipitation and temperature

Acronyn	Explanation	Units
Precipitation related indices		
P	Total rainfall	mm/a or mm/season
Pxcdd	Maximum number of consecutive days with precipitation < 1mm	d
Pxcwd	Maximum number of consecutive days with precipitation > 1mm	d
Pnl90	Number of events exceeding the long-term 90th percentile of precipitation	d
Pfl90	Fraction of total precipitation from events > long-term 90th percentile of precipitation	%
Pint	Mean precipitation on wet days (days with precipitation >1 mm)	mm/d
Px5d	Maximum total precipitation from any consecutive 5 days	mm/5d
Temperature related indices		
Txav	Mean of daily maximum temperature	°C
Tnav	Mean of daily minimum temperature	°C
Trav	Mean of daily diurnal temperature range	°C
Txq90	90th percentile of daily maximum temperature (hot days)	°C
Tnq10	10th percentile of daily minimum temperature (cold nights)	°C
Tnfd	Number of frost days Tmin<0°C	d
Txf90	% days Tmax > long term mean 90th percentile	%
Tnf10	% days Tmin < long term mean 10th percentile	%

3.2.4. Trend estimation

A trend analysis was performed on the time series of the 15 indices using the non-parametric Mann–Kendall (MK) statistical test (Mann 1945; Kendall 1975). This test allows us to investigate long-term trends of data without assuming any particular distribution. Moreover, it is less influenced by outliers in the data set as it is non-parametric. Statistical significance of the trends is evaluated at the 10% level of significance against the null hypothesis that there is no trend in the analyzed variable. The test statistic S of the MK test is defined as follows:

$$S = \sum_{i=1}^{n-1} \sum_{j=i+1}^{n} sgn(x_j - x_i) \qquad (3.1)$$

$$sgn(x_j - x_i) = \begin{cases} +1, x_j > x_i \\ 0, \quad x_j = x_i \\ -1, x_j < x_i \end{cases} \qquad (3.2)$$

where n is the data record length, x_i and x_j are the sequential data values. For $n \geq 10$, the test statistic S is approximately normally distributed with the mean and variance given by

$$E[S] = 0$$
$$Var(S) = \frac{n(n-1)(2n+5) - \sum_{i=1}^{m} t_i(t_i-1)(2t_i+5)}{18} \qquad (3.3)$$

where m is the number of tied groups and t_i is the size of the ith tied group. The standardized test statistics Z is computed by

$$Z = \begin{cases} \frac{S-1}{\sqrt{Var(S)}} & S > 0 \\ 0 & S = 0 \\ \frac{S+1}{\sqrt{Var(S)}} & S < 0 \end{cases} \qquad (3.4)$$

The standardized MK test statistics Z follows the standard normal distribution with a mean of zero and variance of one under the null hypothesis of no trend. If $|Z| > Z_{1-\alpha/2}$, the null hypothesis is rejected at α level of significance. A positive value of Z indicates an upward trend and whereas a negative value indicates a downward trend.

Serial correlation

The Mann-Kendall approach requires the data to be serially independent (von Storch and Navarra 1995). The presence of serial correlation in the analyzed time series can have serious impacts on the results of a trend test. A positive serial correlation can overestimate the probability of a trend and a negative correlation may cause its underestimation. In this study, Mann-Kendall test was used in conjunction with the widely used method of pre-whitening. The pre-whitening removes serial correlation from the data by means of the following formula:

$$y_t = x_t - \emptyset x_{t-1} \qquad (3.5)$$

where y_t is the pre-whitened time series value, x_t is the original time series value for time interval t, and $\emptyset$ is the lag 1 autocorrelation coefficient.

In this study, the pre-whitening was applied when a serial dependence was found significant at the 5% level. In our case, most of the studied variables did not show significant serial correlation except the annual mean minimum temperature series at Henan, Hongyuan,

Xinghai, Jiuzhi and Zeku stations. Before applying the trend test, the pre-whitening was applied to remove serial correlation from these time series.

Theil and Sen's median slope estimator

The Theil-Sen estimator is used to estimate the slope of linear trends (Sen 1968). The estimator is also termed 'median of pair-wise slopes'. It is frequently applied in climatological practice (e.g. Kunkel et al. 1999; Zhang et al. 2001) and outperforms the least-squares regression in computing the magnitude of linear trends when the sample size is large (Zhang et al. 2004).

The slope estimates of N pairs of data are first computed by

$$Q_i = (x_j - x_k)/(j - k) \qquad \text{For } i = 1, \dots N \qquad (3.6)$$

where x_j and x_k are data values at times j and k ($j > k$), respectively. The median of these N values of Q_i is the Sen's estimator of slope.

3.3. Results

The results of the Mann–Kendall test are summarized in Table 3.2 for temperature and in Table 3.3 for rainfall. Presented is the percentage of stations with significant negative trend, significant positive trend and no trend or insignificant trend for each of the indices and for the three study periods. The spatial patterns of the temperature indices are displayed for the longer period 1961–2006, while for rainfall it is done for the period 1960–2000 in order to seek a balance between the spatial coverage and the observations period length. Figure 3.1 shows the trend sign and the change per decade for the eight annual temperature indices. Figures 3.2, 3.3, 3.4, 3.5 present similar data as in Figure 3. 1, but for winter, spring, summer and autumn, respectively. A similar order presents the results for the rainfall (Figures 3.6, 3.7, 3.8, 3.9, 3.10).

Table 3.2: Percentage of stations with significant negative trend (-), significant positive trend (+) and no trend or insignificant trend (0) in temperature indices at the 10% level.

Season	Indices	1960-1990 (13 stations)			1960-2000 (12 stations)			1960-2006 (10 stations)		
		−	0	+	−	0	+	−	0	+
DJF	Txav	0	100	0	8.3	91.7	0	0	70	30
	Tnav	0	23.1	76.9	8.3	25	66.7	10	10	80
	Trav	53.8	46.2	0	58.3	25	16.7	60	20	20
	Tnfd	0	100	0	0	100	0	0	100	0
	Txq90	0	76.9	23.1	0	75	25	0	60	40
	Tnq10	0	46.2	53.8	16.7	25	58.3	10	20	70
	Txf90	0	69.2	30.8	0	75	25	0	60	40
	Tnf10	76.9	23.1	0	58.3	33.4	8.3	70	20	10
MAM	Txav	30.8	69.2	0	8.3	83.4	8.3	10	80	10
	Tnav	0	46.2	53.8	8.3	22.5	69.2	10	10	80
	Trav	76.9	23.1	0	75	16.7	8.3	70	20	10
	Tnfd	0	92.3	7.7	16.7	75	8.3	20	70	10
	Txq90	0	100	0	8.3	83.4	8.3	10	90	0
	Tnq10	0	53.8	46.2	0	33.3	66.7	10	20	70
	Txf90	7.7	92.3	0	8.3	91.7	0	10	90	0
	Tnf10	53.8	46.2	0	66.7	25	8.3	70	20	10
JJA	Txav	0	100	0	0	75	25	0	40	60
	Tnav	0	69.2	30.8	8.3	33.4	58.3	0	20	80
	Trav	23.1	76.9	0	16.7	83.3	0	20	80	0
	Tnfd	38.5	61.5	0	58.3	33.4	8.3	80	10	10
	Txq90	0	100	0	0	75	25	0	30	70
	Tnq10	0	46.2	53.8	16.7	16.6	66.7	0	30	70
	Txf90	0	100	0	0	66.7	33.3	0	30	70
	Tnf10	46.2	46.1	7.7	66.7	16.6	16.7	60	40	0
SON	Txav	7.7	92.3	0	8.3	66.7	25	0	20	80
	Tnav	7.7	61.5	30.8	8.3	41.7	50	10	30	60
	Trav	0	100	0	8.3	83.4	8.3	10	70	20
	Tnfd	0	92.3	7.7	0	91.7	8.3	10	80	10
	Txq90	7.1	92.9	0	0	100	0	0	90	10
	Tnq10	7.7	46.1	46.2	8.3	41.7	50	10	20	70
	Txf90	7.7	92.3	0	0	91.7	8.3	0	60	40
	Tnf10	38.5	53.8	7.7	66.7	25	8.3	60	30	10
Annual	Txav	7.7	92.3	0	8.3	75	16.7	10	40	50
	Tnav	7.7	23.1	69.2	8.3	25	66.7	10	10	80
	Trav	61.5	38.5	0	58.3	25	16.7	60	10	30
	Tnfd	23.1	69.2	7.7	33.3	58.4	8.3	70	20	10
	Txq90	7.7	92.3	0	8.3	91.7	0	0	60	40
	Tnq10	0	46.2	53.8	8.3	33.4	58.3	10	20	70
	Txf90	0	92.3	7.7	0	83.3	16.7	0	30	70
	Tnf10	61.5	30.8	7.7	66.7	25	8.3	80	10	10

Table 3.3: Percentage of stations with significant negative trend (-), significant positive trend (+) and no trend or insignificant trend (0) in rainfall indices at the 10% level.

Season	Indices	1960-1990 (14 stations)			1960-2000 (10 stations)			1960-2006 (7 stations)		
		−	0	+	−	0	+	−	0	+
DJF	Px5d	0	85.7	14.3	0	90	10	0	100	0
	Pxcdd	7.1	92.9	0	30	70	0	28.6	71.4	0
	Pxcwd	0	100	0	0	90	10	0	100	0
	Pint	0	100	0	0	100	0	0	100	0
	Pfl90	0	92.9	7.1	0	90	10	0	100	0
	Pnl90	0	92.9	7.1	0	80	20	0	100	0
	P	0	71.4	28.6	0	70	30	0	71.4	28.6
MAM	Px5d	0	100	0	0	100	0	0	100	0
	Pxcdd	7.1	92.9	0	10	90	0	0	100	0
	Pxcwd	0	100	0	0	100	0	0	100	0
	Pint	0	100	0	0	100	0	0	85.7	14.3
	Pfl90	0	100	0	0	100	0	14.3	85.7	0
	Pnl90	0	100	0	0	90	10	0	100	0
	P	0	92.9	7.1	0	80	20	0	71.4	28.6
JJA	Px5d	0	100	0	0	90	10	0	100	0
	Pxcdd	0	100	0	0	100	0	0	100	0
	Pxcwd	0	100	0	0	100	0	14.3	85.7	0
	Pint	0	100	0	0	100	0	0	100	0
	Pfl90	0	100	0	0	100	0	0	100	0
	Pnl90	0	100	0	0	100	0	0	100	0
	P	0	100	0	0	100	0	0	100	0
SON	Px5d	0	100	0	20	80	0	0	100	0
	Pxcdd	0	85.7	14.3	0	70	30	0	100	0
	Pxcwd	7.1	92.9	0	10	90	0	0	100	0
	Pint	0	100	0	0	100	0	0	100	0
	Pfl90	7.1	92.9	0	10	90	0	0	100	0
	Pnl90	0	100	0	20	80	0	0	100	0
	P	0	100	0	20	80	0	14.3	85.7	0
Annual	Px5d	0	100	0	0	100	0	0	100	0
	Pxcdd	0	100	0	0	100	0	14.3	85.7	0
	Pxcwd	0	100	0	10	90	0	0	100	0
	Pint	0	100	0	0	100	0	0	100	0
	Pfl90	0	100	0	10	90	0	0	100	0
	Pnl90	0	100	0	0	100	0	0	100	0
	P	0	100	0	10	80	10	0	85.7	14.3

3.3.1. Temperature

Overall, significant trends dominate in all the temperature indices for the three different periods. Trends in the longer period 1961–2006 are more pronounced and frequent than in the two shorter periods (1961–1990 and 1961–2000). This behaviour is consistent with the greater power of the trend test for longer analysis periods.

In the period 1961–2006, significant increasing trends dominate in mean daily maximum temperature (*Txav*) on an annual basis with 50% of the stations showing significant positive trends, whereas few stations show an increasing trend in the two shorter periods. On a seasonal basis, *Txav* shows the largest warming trends in autumn with 80% of the stations having a significant positive trend while only 25% show a significant positive trend in the period 1961–2000 and no station shows a significant positive trend in the period 1961–1990. Similarly, significant warming trends also dominate in summer with 60% of the stations having significant positive trends in the period 1961–2006 while there are few or even no stations with significant positive trends in two shorter periods. In winter, 30% of the stations show a significant positive trend in the period 1961–2006 while almost all stations show no significant trend for the two short periods. The mean daily minimum temperature (*Tnav*) shows significant increase in the three periods with the period 1961–2006 having the largest percentage of significant warming trends (80%), followed by 1961–1990 and 1961–2000 with 69.2% and 66.7%, respectively. The annual pattern of *Tnav* is very consistent throughout the year with similar proportions of significant positive trends across the different seasons.

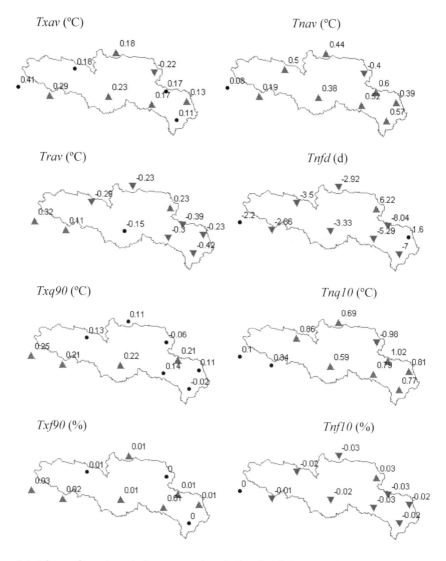

Figure 3.1: Signs of trends and change per decade for the eight annual temperature indices in the study area for the period 1961-2006. Significant increasing (decreasing) trends are marked by filled triangles [▲ (▼)]. Insignificant trends are marked by small dots [•]

A significant decrease has been found in mean diurnal temperature range (*Trav*) on an annual basis. About 60% of the stations exhibited a significant downward trend for the three periods. Spring has the largest proportion of stations (70–77%) showing a significant decrease in the three periods, followed by winter with about 53–60% of the stations showing a significant downward trend. During summer and autumn, the majority of the stations remain stationary, and only few of the stations show significant trends. The number of frost days (*Tnfd*) has significantly declined in the three periods. The period 1961–2006 has the largest number of the stations (70%) showing a significant decline while there is less percentage of the stations (23 and 33%, respectively) with significant negative trends for the periods 1961–1990 and 1961–2000. The largest decline occurs in summer with 80% of the stations showing significant decline in the period 1961–2006. However, no significant change has been detected at the majority of stations in the other seasons.

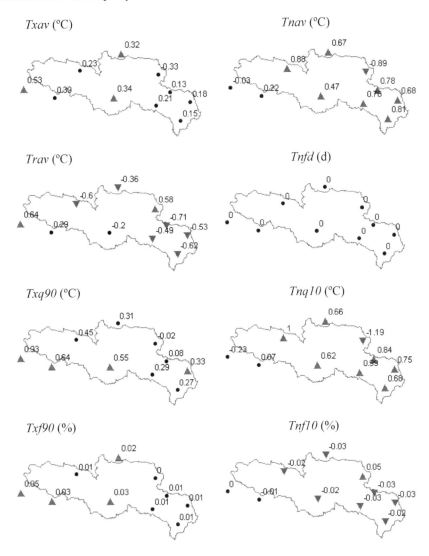

Figure 3.2: Same as in Fig. 3.1, but for winter

A significant decrease has been found in mean diurnal temperature range (*Trav*) on an annual basis. About 60% of the stations exhibited a significant downward trend for the three periods. Spring has the largest proportion of stations (70–77%) showing a significant decrease in the three periods, followed by winter with about 53–60% of the stations showing a significant downward trend. During summer and autumn, the majority of the stations remain stationary, and only few of the stations show significant trends. The number of frost days (*Tnfd*) has significantly declined in the three periods. The period 1961–2006 has the largest number of the stations (70%) showing a significant decline while there is less percentage of the stations (23 and 33%, respectively) with significant negative trends for the periods 1961–1990 and 1961–2000. The largest decline occurs in summer with 80% of the stations showing significant decline in the period 1961–2006. However, no significant change has been detected at the majority of stations in the other seasons.

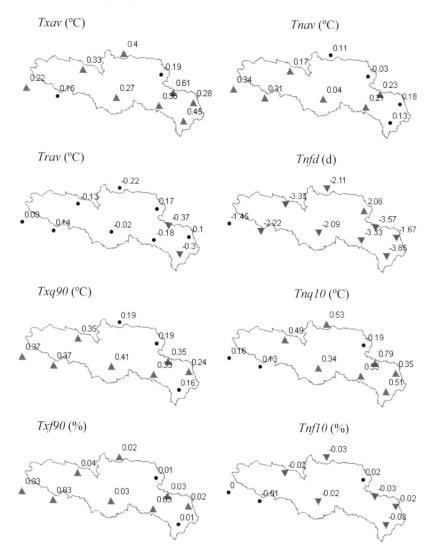

Figure 3.3: Same as in Figure 3. 1, but for summer.

Significant increasing trends also dominate in *Txq90* on annual basis during the period 1961–2006 with 40% of the stations showing a significant positive trend, whereas no significant changes are noted for the other two periods. On a seasonal analysis, summer has the largest number of stations (25% and 70%, respectively) showing a significant upward trend for the two periods 1961–2000 and 1961–2006. Winter follows with 25–40% of the stations having significant positive trends for the three periods. Spring and autumn shows an insignificant change. For *Tnq10*, significant increase has been detected at about 50–70% of the stations in the three periods. The pattern is consistent throughout the year.

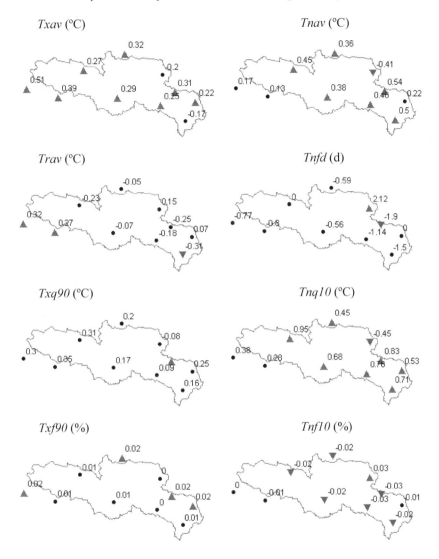

Figure 3.4: Same as in Figure 3. 1, but for autumn.

On an annual basis, there are a large number of stations with increasing trends for *Txf 90* over the period 1961–2006, with as many as 70% of the stations reach significance while no significant change was noted for the two shorter periods. Summer has the largest number of stations (70%) showing significant increasing trends in the period 1961–2006 while there are less (33%) or even no stations with significant positive trends in the periods 1961–2000 and 1961–1990, respectively. Similarly, a significant increase also dominate in winter and autumn with about 40% of the stations showing significant positive trends over the period 1961–2006 while spring shows insignificant changes. In contrast, *Tnf10* is dominated by a significant decrease in the three periods with 60–80% of the stations having significant negative trends. The annual pattern of *Tnf10* is very similar and consistent throughout the different seasons.

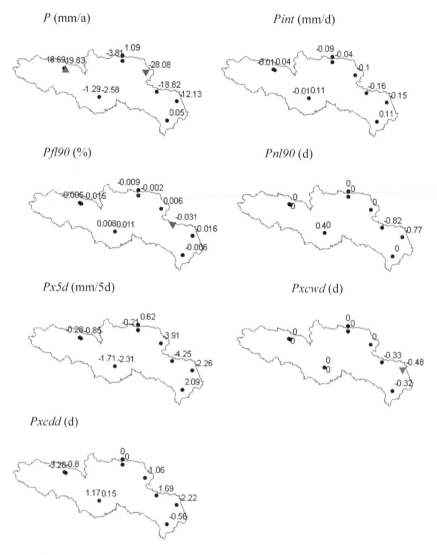

Figure 3.5: Signs of trends and change per decade for the seven annual rainfall indices in the study area for the period 1960-2000. Significant increasing (decreasing) trends are marked by filled triangles [▲ (▼)]. Insignificant trends are marked by small dots [•]

3.3.2. Rainfall

In contrast to the temperature indices, there are no significant changes in all the rainfall indices over the study region for the three study periods. This is indicated by no significant trends being detected at 70–100% of the stations. Trends in the two longer periods (1960–

2000 and 1960–2006) are more consistent and pronounced than the shorter period (1960–1990). Although in the majority of the stations no significant trends have been detected for the seven rainfall indices, some seasonal and spatial differences are noticed.

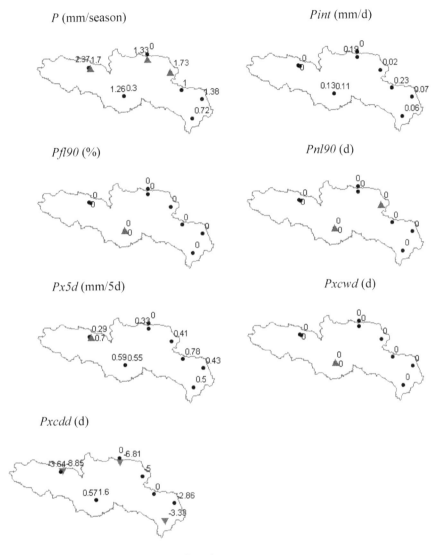

Figure 3.6: Same as in Figure 3. 5, but for winter

No significant changes are noted in annual rainfall total (P) for the three periods except in the upper part of the study region where P has significantly increased at a rate of 19.83 mm/decade. Insignificant change in annual rainfall is also reported by Zhao et al. (2007) and Xu et al. (2007) for the Yellow River source region over the period 1960–2000. Winter has the largest proportion of stations (about 28.6–30%) showing a significant increase in P, followed by spring with 7.1–28.6% of stations showing a significant positive trend. As in the annual basis, P shows a significant increasing trend in winter and spring in the upper part of the study region. In contrast, 14.3–20% of stations exhibit significant decrease in P in autumn while all stations show no significant change in summer. The mean precipitation per wet day (*Pint*) is characterized on an annual basis by no significant trends being detected at all stations. The annual pattern of *Pint* is consistent throughout the different seasons.

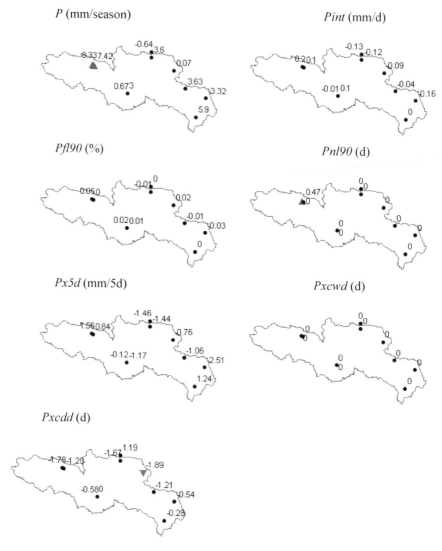

Figure 3.7: Same as in Figure 3. 5, but for spring.

At the majority of stations (85.7% of the stations) no significant trends were found in *Pfl90* for the three periods on both annual and seasonal basis. Similarly, the number of events exceeding the long-term 90th percentile of precipitation (*Pnl90*) also shows no significant trends at the majority of stations (80%). The index of accumulated rainfall during the 5 days with heaviest rainfall (*Px5d*) is related to the most intense rainfall events. On annual basis, Px5d tends to be stationary at all stations. In winter, significant increase in *Px5d* occurs in the upper part of the study region, whereas 20% of the stations, mainly in the lower part of the study region, show significant decrease in autumn.

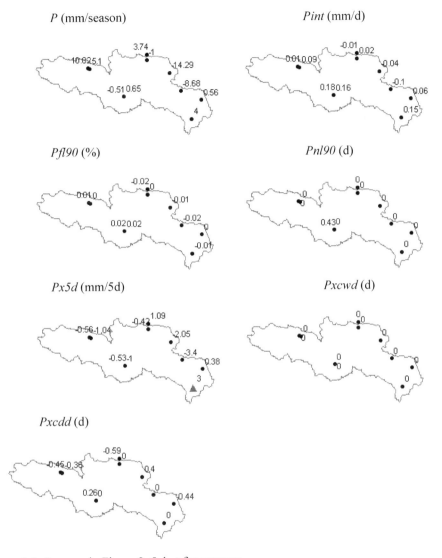

Figure 3.8: Same as in Figure 3. 5, but for summer.

Almost all stations show no significant trends in the length of wet periods (*Pxcwd*) on annual basis. The annual pattern is consistent throughout the different seasons. Although no clear pattern is found for *Pxcdd* on annual basis, some seasonal differences are noticed. 7.1–30% of the stations exhibit significant decreasing trend in winter and spring while 14.3–30% show significant increasing trend in autumn.

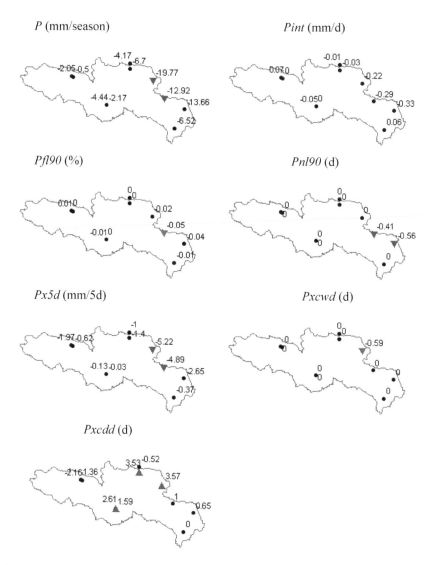

Figure 3.9: Same as in Figure 3. 5, but for autumn.

3.4. Regional average index series

To obtain a general picture of changes in the temperature and rainfall indices over the whole study area, the trend test is also carried out for the regional average index series. Table 3.4 presents the trend test results for regionally averaged indices for the period 1960–2006. The time series of annual region averaged temperature and rainfall indices can be seen in Figures. 3.10 and 3.11, respectively.

3.4.1. Temperature

On annual basis, significant trends dominate all the regionally averaged temperature indices. Out of eight indices, five (*Txav,Tnav, Txq90, Tnq10* and *Txf90*) show a significant upward trend while the remaining three (*Trav, Tnfd* and *Tnf10*) show a significant downward trend. In winter, all the indices show a significant trend except *Tnfd* with no significant change. Both *Trav* and *Tnf10* exhibit a significant downward trend while the remaining five indices show a significant upward trend. In comparison to winter, spring has less number of indices with significant trends. *Tnav* and *Tnq10* show a significant upward trend while *Trav* and *Tnf10* show a significant downward trend. In summer, all the indices show a significant trend except *Trav*. *Tnfd* and *Tnf10* show a significant downward trend while the contrast occurs for the remaining indices. In autumn, four indices (*Txav, Tnav, Txq90* and *Tnq10*) show a significant upward trend, and one index (*Tnf10*) shows a significant downward trend. In spite of some seasonal differences, it is noticeable that throughout the year both *Tnav* and *Tnq10* show a significant upward trend while *Tnf10* shows a significant downward trend. This indicated that significant warming occurs in the minimum temperature related indices throughout the year.

Table 3.4: Mann–Kendall statistics for regional average indices series

	DJF	MAM	JJA	SON	Annual
Temperature indices					
Txav	1.75	0.13	2.63	3.07	2.94
Tnav	3.84	4.69	3.83	2.73	5.15
Trav	-2.00	-3.12	-1.11	-0.89	-2.42
Tnfd	0.13	-0.49	-4.10	-0.53	-2.91
Txq90	2.63	0.51	2.78	0.89	1.83
Tnq10	2.76	3.23	4.32	4.16	4.39
Txf90	2.31	0.44	3.57	2.37	3.62
Tnf10	-3.19	-3.59	-4.15	-3.38	-4.95
Rainfall indices					
Px5d	2.01	-0.31	-0.87	-0.79	-0.89
Pxcdd	-0.15	-1.71	0.24	0.73	-1.12
Pxcwd	1.60	1.22	-1.44	-0.77	-1.48
Pint	0.30	-2.16	-0.81	-1.02	-2.03
Pfl90	0.07	-1.82	-1.90	-1.10	-2.07
Pnl90	0.36	-0.71	-1.90	-1.34	-1.62
P	2.70	1.91	-0.50	-1.12	-0.29

Numbers rendered in bold and italics indicate significance at the 10% level

3.4.2. Rainfall

On annual basis, both the rainfall intensity (*Pint*) and the contribution of moderately heavy rainfall events to total *P* (*Pfl90*) show a significant decreasing trend, whereas no significant changes are found for the remaining indices. Seasonal precipitation shows a significant increasing trend in winter, which is accompanied by a significant increasing trend in the maximum 5-d rainfall (*Px5d*). In spring, there is a significant increasing trend in precipitation (*P*) and a significant decreasing trend in the duration of dry spells (*Pxcdd*), rainfall intensity (*Pint*) and the contribution of moderately heavy rainfall events to total *P* (*Pfl90*). A significant decreasing trend in the frequency and contribution of moderately heavy rainfall events to total *P* (*Pfl90, Pnl90*) is observed in summer. However, none of the indices show significant trends in autumn.

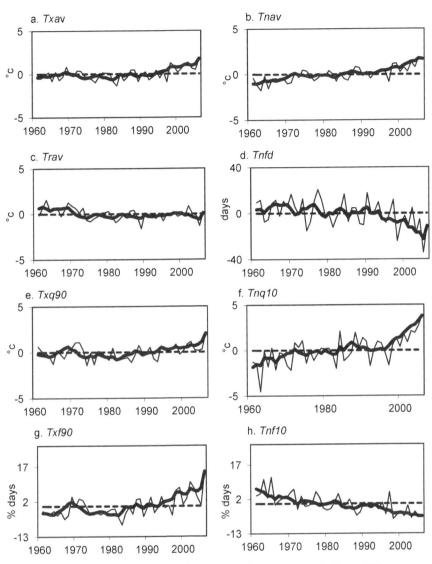

Figure 3.10: Time series for annual region averaged temperature indices. Thin curves show the regional average. Thick curves show 5 years moving average. Horizontal dashed lines show the 1961-2006 average. Data are plotted as anomalies from the 1961-2006 average.

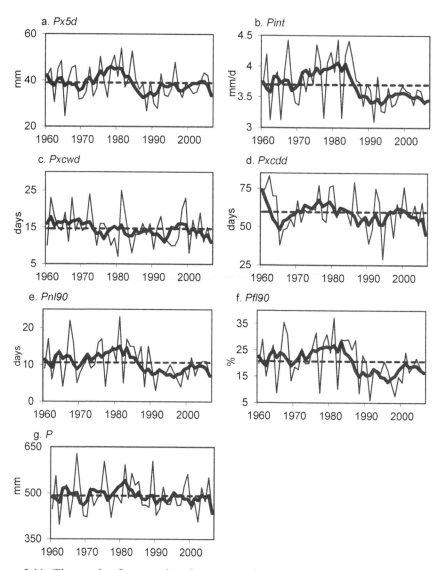

Figure 3.11: Time series for annual region averaged rainfall indices. Thin curves show the regional average. Thick curves show 5 years moving average. Horizontal dashed lines show the 1960-2006 average

3.5. Discussion and conclusions

In this study we analyzed spatio-temporal changes in a set of daily rainfall and temperature indices on both annual and seasonal basis for the Yellow River source region over the three periods: 1960–1990, 1960–2000 and 1960–2006. Changes in the daily data values and spatio-temporal distribution of rainfall and temperature have important implications for water supply in the whole basin that is mainly limited by the water availability in this region.

Significant warming trends dominate the study region over the second half of the twentieth century as a whole. The warming in the study region mainly results from significant increase in winter minimum temperature. This finding is in agreement with results for the whole Yellow River basin (Fu et al., 2004; Zhang et al., 2008). Furthermore, increase in the minimum temperature is much larger than that in the maximum temperature, which results in significant reduction in the diurnal temperature range over most of the region except in the west part of the region where the contrast occurs. This reduction is particularly strong in winter with some stations having trends as much as a $0.5{\sim}0.7°C$ decade^{-1} decrease. The reduction in the diurnal temperature range over most of this region is also reported by Tang et al. (2008) for the period 1960–2000. This is consistent with the trend in the global diurnal temperature range (Easterling et al., 1997). There is a widespread decrease in the number of annual frost days with the largest reduction in summer. On annual basis, the frequency and magnitude of hot events show an upward trend over most of the region with some notable seasonal differences. For example, summer shows the largest increase in both the frequency and magnitude of hot events, with winter to follow, whereas spring and autumn show no significant change. Compared to the annual pattern in the hot events, changes in the magnitude and the frequency of cold events is very consistent throughout the year, e.g. the magnitude of the cold events has significantly increased while the frequency showed a decreasing trend. Zhang et al. (2008) also indicated that the upper reach of the Yellow River is characterized by a significant upward trend of frequency of extreme hot events and by a significant downward trend of frequency of extreme cold events.

Changes in all the indices are spatially coherent for all the stations except for Henan station located in the eastern part of study, which consistently shows opposite trends compared to other stations. Similar findings for the mean air temperature have been reported by Zhao et al. (2007) and Xu et al. (2007). They also found the mean air temperature has decreased in the area around Henan station. This behavior may indicate large climate variability in the mountainous region. However, this hypothesis has not yet been verified. Although changes are spatially coherent for most indices, the rate of change varied, e.g. changes in the cold indices for the western region is much smaller than that for the rest parts of region while the contrast occurs in the warm indices.

Although most stations across the study area did not show any significant changes in the rainfall indices, some noticeable changes were observed for the regionally averaged indices: (a) On annual basis, there is a significant decline in average rainfall intensity and contribution of moderately heavy rainfall events to total P across the study area. (b) Winter rainfall has generally increased significantly, which is accompanied by a significant increase in $Px5d$. (c) Spring rainfall is also found to have a significant increasing trend, which is accompanied by a significant decline in the number of dry days, average rainfall intensity and contribution of moderately heavy rainfall events to total P. (d) Both the frequency and contribution of moderately heavy rainfall events to total P has significantly decreased in summer.

The results of this study indicate that the climate in the Yellow River source region has become warmer and experienced some seasonally varying changes in rainfall, which is in agreement with the conclusions by Niu et al. (2004) for the northeastern Tibetan Plateau. In addition, it partly supports an emerging global picture of warming and prevailing positive trends in rainfall extremes over the mid-latitudinal land areas of the Northern Hemisphere in winter. However, this study is based on relatively short periods (40–45 year) and rather sparse station coverage. Thus, it is unclear as to whether these trends are part of a longer period of oscillation or the result of long term climate change.

4. Streamflow trends and climate linkages in the YRSR[3]

Abstract: Much of the discussion on hydrological trends and variability in the source region of the Yellow River centres on the mean values of the mainstream flows. Changes in hydrological extremes in the mainstream as well as in the tributary flows are largely unexplored. Although decreasing water availability has been noted, the nature of those changes is less explored. Here we investigate trends and variability in the hydrological regimes (both mean values and extreme events) and their links with the local climate in the source region of the Yellow River over the last 50 years (1959–2008). This large catchment is relatively undisturbed by anthropogenic influences such as abstraction and impoundments, enabling the characterization of largely natural, climate-driven trends. A total of 27 hydrological variables were used as indicators for the analysis. Streamflow records from six major headwater catchments and climatic data from seven stations were studied. The trend results vary considerably from one river basin to another, and become more accentuated with longer time period. Overall, the source region of the Yellow River is characterized by an overall tendency towards decreasing water availability. Noteworthy are strong decreasing trends in the winter (dry season) monthly flows of January to March and September as well as in annual mean flow, annual 1-, 3-, 7-, 30- and 90-day maxima and minima flows for Maqu and Tangnag catchments over the period 1959–2008. The hydrological variables studied are closely related to precipitation in the wet season (June, July, August and September), indicating that the widespread decrease in wet season precipitation is expected to be associated with significant decrease in streamflow. To conclude, decreasing precipitation, particularly in the wet season, along with increasing temperature can be associated with pronounced decrease in water resources, posing a significant challenge to downstream water uses.

4.1. Introduction

It is predicted that climate change will lead to an intensification of the global hydrological cycle and can have major impacts on regional water resources (Arnell, 1999; Milly et al., 2008). Such impacts may include the alteration in the magnitude and timing of runoff, frequency and intensity of floods and droughts, and regional water availability. Worldwide, a number of studies have been undertaken to characterize variability and trends in observed records of streamflow and to establish linkages between the atmospheric circulation, climate and streamflow. These include studies in the USA (Fu et al., 2009), Canada (Zhang et al., 2001; Burn and Hag Elnur, 2002; Abdul Aziz and Burn, 2006; Burn, 2008), UK (Hannaford and Marsh, 2006, 2008), Europe (Mudelsee et al., 2003; Tu, 2006), Sweden (Lindstrom and Bergstrom, 2004), Switzerland (Birsan et al., 2005), the Amazon basin in Brazil (Marengo, 2009), the Blue Nile basin (Di Baldassarre et al.,2011), Turkey (Karab¨ork and Kahya, 2009), Iran (Masih et al., 2010), South Korea (Bae et al., 2008) and Southern Africa (Fanta et al., 2001).

In China, Zhang et al. (2005) analysed precipitation, temperature and discharge records from 1951 to 2002 in the Yangtze River Basin, and found an increasing trend in floods in the middle and lower Yangtze basin. They also found similarities in trends and patterns between hydrological variables and meteorological variables. Chen et al. (2006) investigated trends in

[3] This chapter is based on paper Streamflow trends and climate linkages in the source region of the Yellow River, China by Hu, Y., Maskey, S., Uhlenbrook, S. and Zhao, H. 2011. Hydrological Processes 25: 3399-3411. DOI: 10.1002/hyp.8069.

the discharge, temperature and precipitation data in the last 50 years in the Tarim River Basin in northwestern China and suggested that climate change has resulted in streamflow changes in the headwaters of the Tarim River. Fu et al. (2004) studied the hydro-climatic trends of the whole Yellow River Basin for the last 50 years and indicated decrease in the annual natural runoff. Tang et al. (2008) examined trends in annual discharge at six hydrological stations located in the mainstream of the Yellow River from 1960 to 2000. They found a significant decreasing trend for all stations except for Tangnag station where no noticeable changes in annual discharge have been detected. Zheng et al. (2007) investigated changes in streamflow regime for four headwater catchments (i.e. Huangheyan, Jimai, Maqu and Tangnag) of the Yellow River basin from 1956 to 2000, and have not found significant trends in streamflow. Wang (2009) studied climate variations in the upper Yellow River and its impact on eco-hydrology for the period 1955–2005, and suggested decreasing annual runoff as well as summer and autumn runoff for all hydrological stations in both the mainstream and the tributary.

Although some of the studies present evidence of significant variability in streamflow, with both increasing and decreasing trends, all these trends cannot be definitively attributed to climatic changes. Anthropogenic effects, such as the construction of large reservoirs or changes in land use, can hinder the ability to detect the impact that climate change may have on water resource systems (Uhlenbrook, 2009). This is particularly the case for the lower and middle Yellow River, where human activities are the dominating factor leading to significant decrease in runoff in the Yellow River basin during the past five decades (Fu et al., 2004; Fu et al., 2007) and a stabilization of low flows because of reservoir management in the last 50 years. However, in this study the source region of the Yellow River was chosen for exploring the trends and variability in hydrological variables based on the following considerations: (1) this region is relatively pristine and has been subject to very few human interventions, which provides a rare opportunity for assessing trends and natural variability on the flow regime without direct human impacts in a large river basin (121 972 km^2), (2) water availability in this region has important implications for water supply in the entire Yellow River basin as it contributes about 35% of the total annual runoff of the whole Yellow River (Zheng et al., 2007) and (3) previous work indicates that the Tibetan Plateau, where the study area is located, is very sensitive to global climate change (Liu and Chen, 2000). Furthermore, comparatively little research has been conducted on trends and variability in hydrological extremes in the source region of the Yellow River partially due to the lack of sufficient data in this remote region. Trends in temperature and precipitation extremes in the source region of Yellow River were investigated by Hu et al. (2010) for the period 1960–2006. But, their study did not include trends in the hydrological regimes. Although Zheng et al. (2007) studied changes in the streamflow regime in the source region of the Yellow River from the 1950s to 2000, their focus was on changes in the mean streamflows in the four mainstream catchments, and they did not include changes in the hydrological extremes and streamflow regimes of the tributary as well as the linkage between climate and streamflows. Lan et al. (2010) examined the response of runoff in the source region of the Yellow River to climate warming from 1960 to 2002 with a focus on decadal variations in the mean hydro-climatic variables. Liang et al. (2010) analysed the periodicity of precipitation, temperature and discharge record in the source region of the Yellow River from 1955 to 1999 and explored the linkages among them. Their study focused only on the mainstream discharge at the Huangheyan station, and no other station discharges and their climatic linkages were reported.

Overall, previous studies reported in the literature are limited to the mean values of the mainstream flows. There is no comprehensive study on changes in hydrological extremes as well as in the tributary flows. However, evidence from climate models and hydrological studies suggests that extreme events, such as floods and droughts, are likely to change with global warming (Battisti and Naylor, 2009). Climate change could, therefore, result in: (i) increases or decreases in extreme event magnitudes; (ii) changes in the timing of extreme events or (iii) changes in the hydrological processes that lead to extreme events. Furthermore, most of the available studies focused on the period from the late 1950s to the 1990s, and the data set in the recent years were not included. Although one of these studies have included the linkages between streamflows and precipitation and temperature in one of the mainstream catchments, but a comprehensive catchment by catchment study of the YRSR has not been conducted.

This article fills this gap by including changes in hydrological regimes of both the mean and the extreme events as well as of both the mainstream and tributaries. Further, in this study we use daily streamflow data for the period 1959 to 2008 which complements the previous work by including the more recent data set in the analysis. We also investigate the relationship between hydrological variables and climatic variables among different catchments in order to better understand the observed hydrological trends and variability.

4.2. Data and methods

Daily streamflow data from six gauging stations operated by Yellow River Conservancy Commission (YRCC) were used in this study. Of the six stations, the four stations namely Huangheyan, Jimai, Maqu and Tangnag are located along the main stream from upstream to downstream, and the two stations namely Tangke and Dashui are located in the tributaries Bai and Hei Rivers, respectively (Figure 2.1; Table 2.2). Daily streamflow records span the period 1959–2008 and the period 1984–2006 for stations along the mainstream and in the tributary, respectively. Further information regarding the streamflow gauging stations is summarized in Table 2.2. Daily precipitation totals measured at seven stations (four of them together with the hydrological stations namely Huangheyan, Jimai, Maqu and Tangnag and three climatic stations namely Jiuzhi, Hongyuan and Ruoergai at other locations as shown in Figure 2.1), and daily mean, maximum and minimum temperature at seven climatic stations (Xinghai, Maduo, Dari, Ruoergai, Maqu, Hongyuan and Jiuzhi) within or adjacent to the Yellow River were collected from YRCC and China Meteorological Administration. Unfortunately, no other station data are available with sufficient length of data record to be considered for this study. The precipitation and temperature data span the periods 1960–2006 and 1961–2006, respectively. Climatic stations within or closest to the drainage area for a streamflow gauging station were assigned to the catchment to examine the relationship between hydrological variables and climatic variables.

Before the analyses, the daily time series were checked for completeness and validated to identify and rectify sequences of anomalous flows. There is discontinuity in the streamflow data for the Huangheyan station in the period from 1968 to 1975. Therefore, years with missing data at Huangheyan station were excluded from the study. Trend tests were applied for three periods. The first period is common to the four mainstream stations starting in 1959 and ending in 1998, and is referred to as the unregulated period. This period is selected to avoid the possible effects of the hydropower plant located in the upstream of Huangheyan station on the natural variability of the streamflow. As discussed in Section on Study Area, the construction of the Huangheyan hydropower plant in 1998 is expected to have some effects on the streamflow at Huangheyan and Jimai stations, while its effect on other downstream stations is assumed negligible as the annual flow at Huangheyan station accounts for less than 5% of downstream stations (Maqu and Tangnag). Therefore, the second period 1959–2008 is used to analyse streamflow trends in the longer periods only at Maqu and Tangnag stations. The third period 1984–2006 is applied to analyse streamflow trends in the tributary stations Tangke and Dashui. Both the tributary stations are unregulated.

4.2.1. Selection of hydrological variables

Hydrological variables reflecting important components of the hydrological regime were selected for analysis. A total of 27 hydrological variables were selected (Table 4.1). These variables can be classified into four groups: (1) mean flows (annual and monthly), (2) maxima/minima for given durations, (3) dates of occurrences of the annual maximum/minimum flows and (4) high- and extreme-flow days. Many of these variables have been used in previous studies (Burn and Hag Elnur, 2002; Gibson et al., 2005; Abdul Aziz and Burn, 2006; Novotny et al., 2006; Masih et al., 2010) and are considered more likely to be sensitive to climate change. The variables associated with minimum flow (i.e. annual minimum flow and timing of annual minimum flow) were evaluated on a water year (1 July to 30 June as commonly used in the Yellow River) basis to limit the effects of dependence between low-flow events, which frequently occur over the transition between calendar years, while a calendar year was considered for all other variables. Following Novotny et al. (2006), high- and extreme-flow days were calculated for each year by counting the number of days the flow rate was above the mean plus one standard deviation and the mean plus two standard deviations, respectively. For the mainstream stations, the mean and the standard deviation were calculated over the period 1959–1998, while they were calculated over the period 1984–2006 for the tributary stations.

4.2.2. Trend and correlation analysis

The time series of all the hydrological variables were analysed using the Mann–Kendall (MK) nonparametric test for trend (Mann, 1945; Kendall, 1975). Statistical significance of the trends is evaluated at the 10% level of significance against the null hypothesis that there is no trend in the analysed variable. To limit the influence of serial correlation on the Mann–Kendall test, several approaches have been suggested for removing the serial correlation from a data set before applying a trend test. However, in our case, none of the data series for detecting trend has significant serial correlation at a 5% level. As an example, the serial correlation analysis results for annual maximum flow at Jimai station and January flow at Tangnag station are shown in Figure 4.1.

The main objective of the correlation analysis is to determine whether the trends in streamflows are attributable to climate change/variability. Similar studies applied in other basins for linking streamflow trends to climatic change were reported by Burn and Hag Elnur (2002), Abdul Aziz and Burn (2006), Burn (2008), Masih et al. (2010) and Novotny et al. (2006). The partial correlation was calculated between hydrological variables and climatic variables using the Pearson method. The calculated correlations were tested for statistical significance at the 10% level. The use of partial correlation results in the identification of the correlation between variables independent of any common trend signal in the two variables (Burn, 2008). Through this mechanism, it is possible to attribute the observed trends in hydrological variables to trends in meteorological variables.

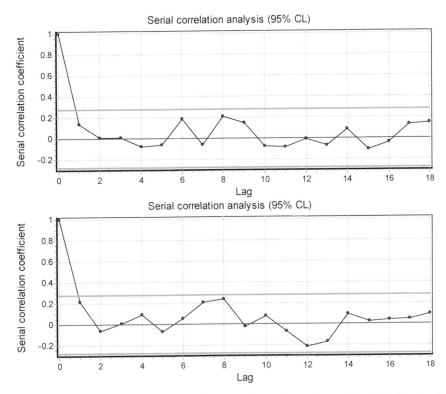

Figure 4.1: The serial correlation analysis for annual maximum flow at Jimai station (top) and January flow at Tangnag station (bottom).

4.3. Results

4.3.1. Trends in hydrological variables

Table 4.1 presents a summary of the results from the trend analysis. Apparent from Table 4.1 are the differences in the results for the different catchments and for the individual hydrological variables. For Huangheyan catchment, no significant trends are present for any of the 27 hydrological variables over the period 1959–1998. Although the trends are not statistically significant, it is interesting that all of them are positive for this catchment except the occurrence of annual maximum flow. Similar to the Huangheyan catchment, there are also no noticeable trends in streamflow variables for the Jimai catchment over the same period, except for the May flow which exhibits a significant increasing trend at 5% level (also shown in Figure 4.2a). The increasing flow in May for this catchment over the period 1956–2005 was also reported by Zheng et al. (2007). In contrast to the Huangheyan and Jimai catchments, a number of hydrological variables show significant decreasing trends in the Maqu catchment over the period 1959–1998. Strong decreasing flows are noted for the months from December to March. As the low flowstypically occur in the winter months, the decreasing flow in the winter months could result in decreasing low flows. As expected, a strong decreasing trend is also found for the low-flow variables, i.e. 1-, 3-, 7-, 30- and 90-day annual minimum flows. As an example, Figure 4.2b and c shows the time series plot for January flow and 7-day annual minimum flow, respectively. However, over the same period the streamflow in the Tangnag catchment showed no significant trends in any of the 27 hydrological variables. Decreases dominate in most of the hydrological variables, although the trends are not statistically significant.

Table 4.1: Trend test results for some hydrological variables, the entries in bold indicates values that are significant at the 10% level.

Variable	Huangheyan 1959-1998	Jimai 1959-1998	Maqu 1959-1998	Maqu 1959-2008	Tangnag 1959-1998	Tangnag 1959-2008	Dashui 1984-2006	Tangke 1984-2006
Annual mean flow	1.02	-0.66	-0.97	**-2.02**	-1.06	**-2.09**	-3.22	-1.58
Monthly mean flows								
January	0.51	-1.18	**-2.97**	**-2.91**	-1.18	**-1.77**	**-1.78**	**2.06**
February	0.45	-0.95	**-2.93**	**-3.24**	-1.20	**-2.17**	**-2.06**	1.41
March	0.52	-0.94	**-2.56**	**-2.87**	-0.71	**-1.82**	**-2.79**	0.81
April	0.11	0.29	0.29	-1.24	0.61	-1.29	**-2.51**	-0.14
May	0.46	**2.17**	1.53	-1.15	1.01	**-1.71**	**-2.31**	0.00
June	0.90	-0.03	0.58	0.19	0.50	-0.18	**-2.45**	**-2.38**
July	1.15	-0.55	-0.76	-1.56	-0.96	-1.56	**-3.49**	-1.63
August	1.02	-0.62	-0.71	-1.30	-0.82	-1.51	**-2.17**	-0.95
September	0.85	-0.73	-1.06	**-1.71**	-1.17	**-1.67**	**-2.54**	**-2.27**
October	0.83	-1.27	-1.43	-1.57	-1.53	-1.30	-0.95	-0.26
November	0.92	-1.04	-1.42	**-1.65**	-1.45	-1.37	-1.11	0.89
December	0.70	-1.11	**-1.87**	-1.31	-0.69	-0.97	-0.90	1.48
Annual maxima								
1-d maxima	1.04	-1.41	-1.45	**-2.39**	-1.25	**-2.59**	**-2.96**	**-2.30**
3-d maxima	1.12	-1.00	-1.40	**-2.35**	-1.13	**-2.44**	**-2.91**	**-2.22**
7-d maxima	1.12	-0.39	-1.24	**-2.30**	-1.06	**-2.32**	**-2.96**	**-2.00**
30-d maxima	1.04	-0.63	-1.38	**-2.11**	-1.22	**-2.17**	**-2.69**	**-2.75**
90-d maxima	1.02	-0.82	-1.15	**-1.97**	-1.20	**-2.07**	**-3.54**	**-2.48**
Annual minima								
1-d minima	0.70	-0.88	**-2.19**	**-2.57**	-1.06	**-1.97**	-1.44	**2.81**
3-d minima	0.64	-0.54	**-2.64**	**-3.10**	-1.20	**-2.16**	-1.47	**2.87**
7-d minima	0.64	-0.67	**-2.87**	**-3.22**	-1.27	**-2.21**	-1.52	**2.93**
30-d minima	0.60	-0.85	**-2.91**	**-3.12**	-1.29	**-2.11**	**-1.75**	**2.42**
90-d minima	0.50	-0.60	**-2.73**	**-2.92**	-1.15	**-1.72**	**-1.97**	**2.42**
Date of maxima	-0.23	-0.70	-0.87	-1.08	-0.34	-0.18	0.34	-1.05
Date of minima	1.04	0.07	-0.62	0.02	0.71	**1.69**	0.74	**-2.02**
High flow days	0.46	-0.71	-0.89	**-1.95**	-1.13	**-2.13**	**-2.86**	**-2.41**
Extreme flow days	0.65	-0.51	-0.77	-1.57	-0.97	**-1.93**	**-3.14**	**-2.64**

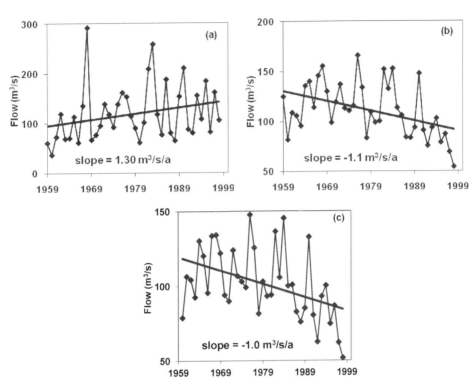

Figure 4.2: Time series plots for (a) May flow at Jimai station, (b) January flow at Maqu station and (c) 7-d annual minimum flow at Maqu station. The symbols show the observed values. Straight black lines show the trend line.

In the second study period (1959–2008), a large number of hydrological variables are noted to have strong decreasing trends for both the Maqu and Tangnag catchments. For the Maqu catchment, annual mean flow shows a clear evidence of strong decreasing trend. Similarly, a strong decreasing trend is also found for both high- and low-flow associated variables. Decreasing trend is noted for high-flow days, while lack of trends is found for extreme-flow days. From the monthly flow variables, January, February, March and November are observed to have significant decreasing trends, whereas the trends are insignificant for the remaining months. Streamflow in the Tangnag catchment displays a similar pattern to that in the Maqu catchment for the period 1959–2008. On an annual basis, a significant decreasing trend was found for annual mean flow as well as high- and low-flow indicators. Both high- and extreme-flow days exhibit a significant decreasing trend. For the date measures, annual minimum flow displays a significant increasing trend as shown in Figure 4.3, implying that annual minimum flow is occurring later in more recent years with the shift from January to February. On a monthly basis, January, February, March, May and September flows are noted to have significant decreasing trends, while the trends are insignificant for the remaining months.

Similar to the main stream stations Maqu and Tangnag, a large number of hydrological variables in the tributary stations Dashui (the Hei River) and Tangke (the Bai River) also exhibit a strong trend over the period 1984–2006. The Hei River is generally characterized by a decrease in water availability. This is apparent in a decrease in the annual mean flow and the monthly flow from January to September. Furthermore, a significant decreasing trend is also noted in high-flow indicators, while a general lack of significant trends for low-flow indicators with the exception of annual 30- and 90-day minimum flows. It is interesting to note that a significant increasing trend is present in the low-flow indicators for the Bai River, while a decreasing trend is found for the high-flow indicators. From the monthly variables, it was found that January flow shows a strong upward trend, while the flows in June and September are observed to have strong downward trends. This coincides with increasing low flows and decreasing high flows. For the timing measures, a strong downward trend is noted in annual minimum flow, while lack of trend is found in annual maximum flow.

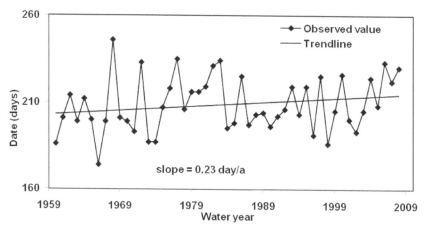

Figure 4.3: The date (starting on July 1th) of annual minimum flow in the Tangnag catchment over the last 50 years.

Comparing the results for the different analysis periods and for the different catchments leads to several observations. An overall decreasing tendency is prevalent in streamflow for the source region of the Yellow River, both for the main stream and for the tributaries. Decrease in streamflow is particularly strong for the two mainstream catchments Maqu and Tangnag over the period 1959–2008, as well as for the two tributaries Bai and Hei Rivers over the period 1984–2006 with many variables reaching the significant level. A number of significant trends were observed for the Tangnag catchment over the period 1959–2008, while no significant trends can be seen over the period 1959–1998.

To examine whether the observed trends over the period 1959–2008 are influenced by multi-decadal variability. The harmonic analysis was applied to annual mean flow of both the Maqu and Tangnag catchments as an example displayed in Table 4.2. It is noted that these two catchments have experienced very similar periodicity. The peaks repeat every 8 years in these two catchments as shown in Figure 4.4. In the last 50 years, annual mean flow experienced about five fluctuations: the 1960s and the 1970s are normal flow periods, the 1980s is a general high-flow period and the 1990s and the 2000s are low-flow periods. These short-term fluctuations will clearly influence the results of trend analyses if study periods begin or end during notably high- or low-flow periods. The trends in the period 1959–2008 are clearly influenced by starting in a normal flow period (1960–1969), which is followed by a normal flow period (1970–1979) and a high-flow period (1980–1989) and ending in a low-flow period (2000–2008). Therefore, apparent trends in the period (1959–2008) for both the Maqu and Tangnag catchments may be influenced by multi-decadal climatic variability rather than climate change.

Table 4.2: Results of periodic changes of both the Maqu and Tangnag catchments.

Catchment	j	T_j	Parameters of harmonics			
			MSD(j)	CP_i	A_j	B_j
Maqu	1	50	2353.8	0.19	-51.81	44.98
	2	8.3	1416.5	0.3	47.78	-23.45
Tangnag	1	50	5207.08	0.19	-75.43	68.73
	2	8.3	3110.75	0.31	72.98	-29.91

Note: j is the j-th harmonic; T_j is the period (year) of the j-th harmonic; MSD(j) is the mean squared deviation of the j-th harmonic; CP_i is the cumulative periodogram; A_j and B_j are the j-th harmonic coefficients.

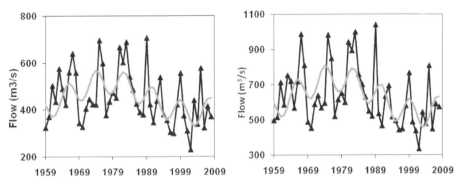

Figure 4.4: Fit of harmonics to the annual mean flows for the Maqu (left) and the Tangnag (right) catchments. The annual mean flows and harmonics are marked by the black triangle line and the gray line, respectively.

4.3.2. Trends in climatic variables

In this study, the climatic data in the YRSR were also investigated for trends over the period 1960–2006 on both annual and monthly basis. For the minimum temperature, strong increasing trends were prevalent in the winter and spring months of November to April as well as in June, while no trends were found in the months of May and July to October. Annual mean minimum temperature, however, showed clear evidence of a strong increasing trend. Similarly, the mean temperature also exhibited a strong increasing trend in the winter and early spring months (November to March) as well as in June and the annual mean temperature. Weak increasing trends were found in the late summer months of July, August and September, while no trends were found in April, May and October. By contrast, no trends were found for the maximum temperature in all months of the year except the months of June and November when almost all stations showed significant increasing trends. However, annual mean maximum temperature displayed a significant increasing trend. Overall, the minimum temperature has the largest increase, followed by the mean and the maximum temperature, respectively, as shown in Figure 4.5.

In majority of the stations, no significant trends were found in monthly and annual precipitation. However, a decline in summer and autumn precipitation and an increase in winter and spring precipitation are noticeable as shown in Figure 4.6. In spite of considerable spatial variability in the trend results for the climatic variables, the overall temperature in the region shows significant warming trend (Figure 4.5). Several other researches also reported dominating warming trends in the region (Xu et al., 2007; Tang et al., 2008; Hu et al., 2010).

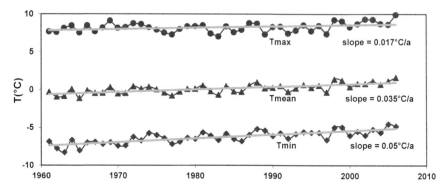

Figure 4.5: Time series plot for areal averaged temperature from 1961-2006. Black curves show the observed values. Gray lines show the linear trend line.

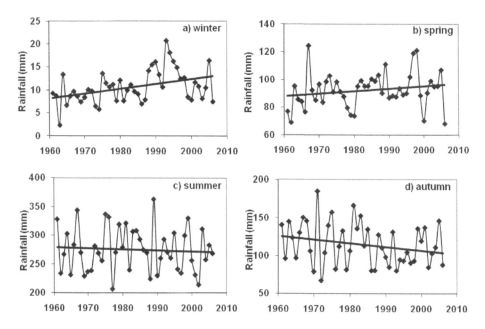

Figure 4.6: Comparisons of seasonal total precipitation.

4.3.3. Streamflow trends and climate linkages

The potential causes of the significant trends in hydrological variables were investigated through a correlation analysis with climatic variables. The correlation was first calculated between each of the climatic variables and each of the hydrological variables. Each of the hydrological variables exhibiting a significant trend was then examined in greater detail to ascertain if there was an explanation for the observed hydrological trend based on the trends in climatic variables exhibiting a relationship with the hydrological variable.

The results of the correlation analysis clearly show that the streamflows are negatively correlated with the mean temperature and positively correlated with the precipitation. Figure 4.7 presents the correlation coefficients between streamflow and arithmetic basin averaged climatic variables on annual basis for the four mainstream catchments. The correlations of the streamflows with precipitation are distinctly increasing from upstream to downstream, whereas no such pattern is appeared with the temperature. Dong et al. (2007) also reported increasing correlation of annual precipitation with annual streamflows from upstream to downstream. They attributed this phenomenon to glacier melting and speculated that glacier melting affects the influences of precipitation on streamflow and this effects decrease from upstream to downstream. However, we suggest an alternative explanation of this phenomenon which might be due to the storage effects of widespread lakes and wetlands in the Huangheyan catchment. As described in Section 2.1, there are about 5300 lakes in the source area of the Yellow River, and about 80% of them are located in the Huangheyan catchment. Precipitation in this catchment mainly contributes to groundwater instead of direct surface water due to the widespread lakes and wetlands (Liang et al., 2007), which results in the weak direct contribution of precipitation to streamflow. In comparison to the Huangheyan catchment, the larger influences of precipitation on streamflow in the other three catchments is probably due to that the effects of lake and wetlands as well as temperature-driven changes in evaporation decrease from upstream to downstream.

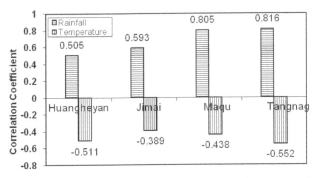

Figure 4.7: Correlation between annual runoff and annual precipitation/temperature for the four main stream catchments.

The correlations between monthly streamflow and precipitation suggest that the mean flow in a month may not be entirely dependent on the precipitation in this month itself, but is often the result of a combined effect of the precipitation in the current and previous months. This mainly results from the occurrence of snowfall in winter which melts in spring to supplement the streamflow and the contribution of runoff from the subsurface storage. For example, May flow in the Jimai catchment is strongly influenced by February to April precipitation ($r = 0.33$), which are significant at the 5% level. May flow in this catchment exhibited a significant increasing trend over the period 1959–1998, while precipitation in the months from February to April showed a strong increasing trend. The positive relationship between the two variables is also apparent from both the data points and the smoothed representations of the series as shown in Figure 4.8a. Therefore, increasing precipitation in the months from February to April could be associated with increasing flow in May. In addition, observed increases in air temperature throughout the winter causing more glacier and snowmelt could also explain this behaviour. Yang et al. (2007) showed a declination of snow cover in the YRSR in the 1990s. Over the same period, the strong positive correlation of January, February, March and low flows with previous September precipitation in the Maqu catchment may imply that decreasing precipitation in September could lead to significant decrease in January, February, March and low flows. As an example, Figure 4.8b and c further demonstrates these relationships between previous September precipitation and January flow and annual 7-day minimum flow, respectively. December flow in the Maqu catchment shows a strong positive correlation with current year July to October precipitation (r ranging from 0.30 in July, 0.38 in August, 0.64 in September to 0.37 in October), which are significant at the 5% level. This relationship is further illustrated by Figure 4.8d, which clearly shows December flow following the similar patterns of July to October precipitation. This similarity in patterns implies that the widespread decrease in precipitation from July to October may be responsible for significant decreasing trends in December flow for the Maqu catchment. Over the period 1959–2008, September flow in both Maqu and Tangnag catchments is found to have a significant positive correlation with precipitation in August (r is about 0.56) and in September (r ranging from 0.63 to 0.7), which indicate that decreased precipitation in August and September may explain significant decrease in September flow.

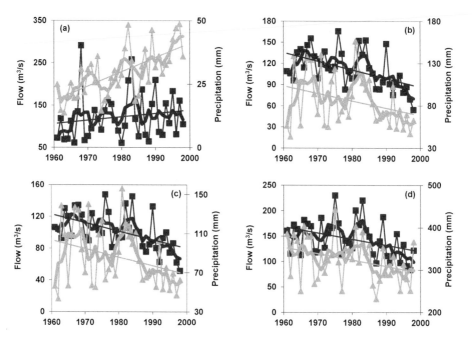

Figure 4.8: Time series plot for (a) May flow and February to April precipitation for the Jimai catchment, (b) January flow and September precipitation, (c) annual 7-d minimum flow and September precipitation, and (d) December flow and July to October precipitation for the Maqu catchment. The observed values for flow and precipitation are marked by the square and the triangle, respectively. The 5-years moving average curves for flow and precipitation are marked by the black and gray thick lines, respectively. The linear trend lines for stream flow and precipitation are marked by black and gray thin lines, respectively.

The correlations of annual mean flow with annual precipitation are particularly strong (>0.8) for Maqu and Tangnag catchments, which are significant at the 0.1% level. The annual mean flow for both the two catchments shows a significant decreasing trend over the period 1959–2008, while annual precipitation shows a decreasing but statistically insignificant tendency. The annual temperature shows a significant increasing trend for the two catchments. It is apparent that annual flow and annual precipitation exhibits similar trends (both decreasing), while annual flow and annual temperature exhibits inverse ones, implying that decrease in annual precipitation along with the increasing temperature may be responsible for significant decrease in annual flow.

In general, the maximum flow variables in all the catchments are closely related to precipitation in the wet season (June, July, August and September) as indicated by a correlation coefficient ranging from 0.2 to 0.63. These correlations are particularly stronger for the Maqu and Tangnag catchments with a correlation coefficient ranging from 0.35 to 0.63. This implies that the widespread decrease in the wet season precipitation might be responsible for significant decreasing trends in annual maximum flow variables over the period 1959–2008. Similarly, the minimum flow variables are also related to the precipitation in the wet season. However, the correlation of the wet season precipitation with the minimum flow is lower ($r < 0.4$ in most cases) in comparison to that with the maximum flow. The date of annual minimum flow in the Tangnag catchment displays a significant increasing trend, implying that annual minimum flow is occurring later in more recent years. This is probably as a result of an earlier snowmelt due to warmer temperature throughout the winter. As mentioned earlier, Yang et al. (2007) showed a declination of snow cover in the YRSR in the 1990s. Ye et al. (2005) noted an earlier snowmelt in the upper Yellow River during the last 50 years. Lu et al. (2009) found that the snowmelt runoff at Tangnag station tended to occur earlier during the period 1957–2000. However, further investigation is needed to get a complete understanding of this issue.

4.4. Conclusions

The study provided an overview of streamflow variability in the YRSR, a relatively pristine, large catchment. Generally, the YRSR is characterized by a decrease in water availability. However, the level of changes differs from catchment to catchment and time period to time period, ranging from little to significant changes in a large number of hydrological variables studied. Over the unregulated period 1959–1998, almost all the hydrological variables in the Huangheyan, Jimai and Tangnag catchments are fairly stable, i.e. no significant changes are seen. In contrast to these catchments, the Maqu catchment displays a strong decreasing trend in the monthly flow from December to March and the low-flow indicators over the same period. Compared to the period 1959–1998, a large number of hydrological variables were noted to have significant decreasing trends over the period 1959–2008 for both Maqu and Tangnag catchments. There is a decreasing trend in the annual mean flow as well as in the annual 1-, 3-, 7-, 30- and 90-day maxima and minima. From the monthly flow variables, January, February, March and September were observed to have strong decreasing trends. For the date measures, annual minimum flow displays a significant increasing trend for the Tangnag catchment, implying annual minimum flow occurred later in the recent years. The later occurrence of annual minimum flow in this catchment appears to be associated to an earlier snowmelt because of increasing temperature in winter.

Among the climatic variables, both the minimum and the mean temperature exhibit an increasing trend in the winter and early spring months of November to April as well as on annual basis. Overall, the region has become warmer with a more significant increase in minimum temperature than in mean and maximum temperature. The warming trend in this region mainly results from increasing minimum temperature in winter. The annual precipitation exhibits a slight decreasing tendency with seasonal differences, e.g. a decrease in the summer as well as the autumn and an increase in the winter as well as the spring.

The correlations between hydrological variables and climatic variables in the four main stream catchments show that the hydrological variables have positive correlations with precipitation but negative ones with temperature. In comparison to temperature, precipitation shows stronger correlation with streamflow in the study region with the exception of the Huangheyan catchment where precipitation and temperature appeared to have equal (but opposite) correlations with streamflow. The latter seems to be the result of temperature-related increases in evaporation for this lakes and wetlands dominated catchment. Furthermore, the correlation of annual precipitation with annual flow increases from upstream to downstream, which is probably a result of the effects of lake and wetlands as well as temperature-related changes in evaporation decrease from upstream to downstream. The similarities in trends and patterns in the hydrological variables and in the climatic variables imply that the trends in hydrological variables may be attributed to changes in climatic variables. Specifically, decreased precipitation in the wet season, along with rising temperature, seems to be responsible for decreased streamflow in the study region. The high flow as well as the low-flow variables are also closely related to precipitation in the wet season (June, July, August and September), indicating that the widespread decrease in the wet season precipitation is expected to be associated with significant decrease in the high flow as well as the low-flow variables over the period 1959–2008.

The hydrological variables studied have experienced the multi-decadal fluctuations as demonstrated by harmonic analysis, implying the observed trends may be influenced by multi-decadal variability. Significant trends in observed streamflow for the study region over the period 1959–2008 may be influenced by starting in a normal flow period and ending in a low-flow period.

5. Downscaling daily precipitation over the YRSR: a comparison of three statistical downscaling methods[4]

Abstract: Three statistical downscaling methods are compared with regard to their ability to downscale summer (June–September) daily precipitation at a network of 14 stations over the Yellow River source region from the NCEP/NCAR reanalysis data with the aim of constructing high-resolution regional precipitation scenarios for impact studies. The methods used are the Statistical Downscaling Model (SDSM), the Generalized LInear Model for daily CLIMate (GLIMCLIM), and the non-homogeneous Hidden Markov Model (NHMM). The methods are compared in terms of several statistics including spatial dependence, wet- and dry spell length distributions and interannual variability. In comparison with other two models, NHMM shows better performance in reproducing the spatial correlation structure, inter-annual variability and magnitude of the observed precipitation. However, it shows difficulty in reproducing observed wet- and dry spell length distributions at some stations. SDSM and GLIMCLIM showed better performance in reproducing the temporal dependence than NHMM. These models are also applied to derive future scenarios for six precipitation indices for the period 2046-2065 using the predictors from two global climate models (GCMs; CGCM3 and ECHAM5) under the IPCC SRES A2, A1B and B1scenarios. There is a strong consensus among two GCMs, three downscaling methods and three emission scenarios in the precipitation change signal. Under the future climate scenarios considered, all parts of the study region would experience increases in rainfall totals and extremes that are statistically significant at most stations. The magnitude of the projected changes is more intense for the SDSM than for other two models, which indicates that climate projection based on results from only one downscaling method should be interpreted with caution. The increase in the magnitude of rainfall totals and extremes is also accompanied by an increase in their inter-annual variability.

5.1. Introduction

It is expected that global climate change will have a strong impact on water resources in many regions of the world (Bates et al., 2008). As a key component of the hydrological cycle, modelling these impacts requires high-resolution regional precipitation scenarios as input to impact models. Currently, global climate models (GCMs) are the most appropriate tools for modelling future global scale climate change. Although the usefulness of these models is unquestionable, they provide information at a resolution that is too coarse to be directly used in impact studies (Xu, 1999). GCMs are also limited in skill to represent subgrid-scale features and dynamics such as convection and topography (Xu, 1999), which are of importance for impact studies on a catchment scale. These limitations become more problematic when a study focuses on precipitation, which strongly depends on subgrid-scale processes (Wilby and Wigley, 2000) and on regions with complex and sharp orography (Schmidli et al., 2006). Consequently, downscaling techniques have been developed to bridge the gap between what,GCMs are able to simulate well and what is needed for the,catchment scale climate change impact research. Among the different downscaling approaches, statistical downscaling is the most widely used one to construct climate change information at a station or local scales because of their relative simplicity and less intensive computation.

[4] This chapter is based on paper Downscaling daily precipitation over the Yellow River source region in China by Hu, Y., Maskey, S. and Uhlenbrook, S. 2013. Theoretical and Applied Climatology 112: 447-460. DOI: 10.1007/s00704-012-0745-4.

Statistical downscaling methods are generally classified into three groups (Wilby and Wigley, 1997): regression models, weather typing schemes and weather generators. Weather generators are adapted for statistical downscaling by conditioning their parameters on large-scale atmospheric predictors (Maraun et al., 2010). They are often used in combination with either regression methods or weather typing schemes. Statistical models can also be divided into single and multi-site methods. The single-site methods model each station independently. The multi-site methods model all sites simultaneously, thereby maintaining inter-station relationships, e.g. spatial correlation, which is one of the important considerations for climate change impact studies over large river basins.

Although there is a large body of literature where an intercomparison of different downscaling methods has been made (e.g. Wilby et al., 1998; Mehrotra et al., 2004; Diaz-Nieto and Wilby, 2005; Frost et al., 2011; Liu et al., 2011, 2012), very few of these studies have dealt with downscaling precipitation in remote mountainous areas. To the best of our knowledge, no other studies have reported downscaling of precipitation in the literature for this catchment with an exception of Xu et al. (2009b), who investigated the response of streamflow to climate change in the headwater catchment of the Yellow River basin using the Statistical Downscaling Model (SDSM) model (Wilby et al., 2002) and a perturbation-based technique called the 'delta-change' method (Prudhomme et al., 2002). The SDSM is a single-site downscaling method. The delta-change method involves adjusting the observed time series by adding the differences (for temperature) or multiplying the ratio (for precipitation) between future and present climates simulated by the GCMs.

The need for regional precipitation scenarios for impact and hydrological studies is particularly urgent for the YRSR. First, located in mountainous areas, this region is expected to be sensitive to global climate change since mountains in many parts of the world (e.g. the Andes and the Himalayas) are very susceptible to a changing climate in view of their complex orography and fragile ecosystem (Beniston, 2003). Second, the Tibetan Plateau, where the study area is located, has been identified as one of the most sensitive areas to global climate change due to its earlier and larger warming trend in comparison to the Northern Hemisphere and the same latitudinal zone in the same period (Liu and Chen, 2000). As a consequence, it can be expected that the YRSR might be particularly susceptible to global climate change. This in turn might have considerable impacts on water availability in the entire Yellow River basin as the source region contributes about 35 % of the total annual runoff of the entire Yellow River. It is therefore important to project future precipitation scenarios over the region in order to provide useful information for impact studies and adaptation/mitigation policy responses. This study is aimed at testing three different statistical downscaling methods in their ability to reconstruct observed daily precipitation over the YRSR and applying them to develop future precipitation scenarios for the region. The three statistical downscaling methods used in this study are: the SDSM (Wilby et al., 2002), the Generalized LInear Model for daily CLIMate (GLIMCLIM) (Chandler, 2002), and the non-homogeneous Hidden Markov Model (NHMM) (Hughes and Guttorp, 1994). All three models have been tested in a range of geographical contexts (Wetterhall et al., 2006; Tryhorn and DeGaetano, 2010; Chandler, 2002; Chandler and Wheater, 2002; Yang et al., 2005; Fealy and Sweeney, 2007; Hughes et al., 1999; Mehrotra et al., 2004; Kioutsioukis et al., 2008). Several previous studies have also compared some of these downscaling models in different river basins (Liu et al., 2011; Frost et al., 2011; Liu et al., 2012). However, all of the previous comparisons were limited to the present climate, and the projections for the future climate were not included in their study.

5.2. Material and methods

5.2.1. Data sets

Observed station data

The observed daily precipitation data used in this study are for the period 1961–1990 from 14 stations distributed throughout the study region. Figure 5.1 depicts the geographical location of the stations in the study region, and Table 2.2 shows their latitude, longitude and altitude. The homogeneity of the data was tested by applying the double mass curve method on a monthly basis for each station (Hu et al., 2012). Slightly less than 0.03 % of the data from two stations were missing, which were infilled using the records from neighbouring stations. We only focus on downscaling summer (monsoon) precipitation because there is negligible rainfall during the remaining part of the year. A threshold of 1 mm/day is used to discriminate between wet and dry days.

Reanalysis data

In addition to the observed data, large-scale atmospheric variables derived from the National Center for Environmental Prediction (NCEP)/National Centre for Atmospheric Research reanalysis data set (Kalnay et al., 1996) on a 2.5°×2.5° grid over the same time period as the observation data were employed for model calibration and validation. These variables include specific humidity, air temperature, zonal and meridional wind speeds at various pressure levels and mean sea level pressure. The predictor domain extends from 30°N to 40°N and from 92.5°E to 107.5°E covering the entire study region.

GCM data

In order to project future precipitation scenarios, output from two GCMs under the Intergovernmental Panel on Climate Change Special Report on Emissions Scenarios A2, A1B and B1 was used: (1) the Canadian Center for Climate Modelling and Analysis 3nd Generation [CGCM3.1 (T47)], and (2) the ECHAM5/MPI-OM GCM from the Max-Planck-Institute for Meteorology, Germany (hereafter ECHAM5). Both models are coupled atmosphere–ocean models. CGCM3 has a horizontal resolution of T47 (approximately 3.75° latitude×3.75° longitude) and 32 vertical levels. ECHAM5 has a horizontal resolution of T63 (approximately 1.875° latitude×1.875° longitude) and 31 vertical levels. These GCM data are obtained from the Program for Climate Model Diagnosis and Intercomparison website (http://www-pcmdi.iinl.gov). The A2, A1B and B1 scenarios span almost the entire IPCC scenario range, with the B1 being close to the low end of the range, the A2 to the high end of the range and A1B to the middle of the range. The GCM simulations corresponding to the present (1961–1990) and future climate (2046–2065) were considered in the analysis. Prior to use in this study, both GCM grids were linearly interpolated to the same 2.5°×2.5° grids fitting the NCEP reanalysis data.

5.2.2. Precipitation indices

Six precipitation indices were selected in order to examine and simulate the changes of the mean and extreme conditions over the study region under future emission scenarios.
1. prcptot—total precipitation (mm);
2. pq95—95th percentile of precipitation on days with precipitation >1 mm (mm/d);
3. pq95tot—total precipitation falling in days with amounts > the corresponding long-term 95th percentile (calculated only for wet days and for the baseline period 1961–1990; mm).

4. pfl95—fraction of total precipitation from events > long-term 95th percentile of precipitation (mm/mm).
5. px5d—maximum total precipitation from any consecutive 5 days (mm).
6. pxcdd—maximum number of consecutive days with precipitation <1 mm (d).

5.2.3. Choice of predictors

There is small agreement on the most appropriate choice of predictor variables. The choice of predictors depends on the region, the characteristics of the large-scale atmospheric circulation, seasonality, the topographic context and the predictand to be downscaled (Anandhi et al., 2008). In this study, the predictors were first selected taking into consideration the monsoon rainfall generation mechanism. Monsoon rainfall in the study region is caused by high temperature in the land area and subsequent generation of low-pressure zone. This results in wind flow withmoisture fromthe Bay of Bengal and the western Pacific Ocean to the land area (Lan et al., 2010), while northwestern cold air current plays a major role in monsoon rainfall generation. Based on this, a number of atmospheric variables were taken as the potential predictors including air temperature, specific humidity, zonal and meridional wind at various pressure levels and mean sea level pressure. These potential predictors were then screened through a correlation analysis with daily monsoon precipitation at each of the 14 stations. Furthermore, experiences and recommendations from similar studies in China and neighbouring regions were also taken into account (Wetterhall et al., 2006; Tripathi et al., 2006; Anandhi et al., 2008; Liu et al., 2011, 2012). The final set of predictors for downscaling of precipitation was selected as follows: specific humidity at 300 and 500 hPa level, zonal wind at 200, 300 and 500 hPa level and meridional wind at 850 and 1,000 hPa level. The explanatory power of a given predictor will vary both spatially and tem-porally for a given predictand. The use of predictors directly overlying the target grid box is likely to fail to capture the strongest correlation (between predictor and predictand) as this domain may be geographically smaller in extent than the circulation domains of the predictors (Wilby and Wigley, 2000). Selecting the spatial domain of the predictors is subjective to the predictor, predictand, season and geographical location (Anandhi et al., 2009). On the basis of these recommendations and monsoon rainfall generation mechanism, the spatial domain of the predictors considered in this study was chosen as 35 grid points lying an extended area covering the entire study region.

The predictors were first standardized at each grid-point by subtracting the mean and dividing by the standard deviation. A principal component analysis was then performed to reduce the dimensionality of the predictors. The first eight principal components, which account for more than 90 % of the total variance, were then used as input to the downscaling model. The principal components were selected on the basis of the percentage of variance of original data explained by individual principal component. This criterion was also used by Tripathi et al. (2006), Anandhi et al. (2008, 2009) and Ghosh (2010). Note that there are also other methods that exist for selecting the principal components, e.g. the elbow method used by Wetterhall et al. (2006).

5.2.4. Statistical downscaling methods

The three downscaling models considered in this study all belong to stochastic downscaling models. They mainly differ in the way their weather generator parameters are conditioned on large-scale predictors or weather states. In SDSM, the multiple linear regression method is used to condition its weather generator parameters on large-scale predictors, whereas in GLIMCLIM and NHMM, this is done using a generalized linear model and a weather state approach, respectively. In addition, SDSM is a single-site model, while GLIMCLIM and

NHMM are multi-site models. Table 5.1 compares the ways the rainfall occurrence, amount and spatial dependence structure are modelled in these three downscaling methods.

Table 5.1: Comparison of the ways the rainfall occurrence, amount and spatial dependence structure are modelled in these three downscaling methods

Notation	SDSM	GLIMCLIM	NHMM
Precipitation occurrence	Depends on current day's predictors and previous day's precipitation occurrence	Depends on current day's predictors and 3 previous days' precipitation occurrence	Depends only on current day's weather state
Precipitation amount	• Using an emprical distribution • Separate paramters for each station and each month	• Using a gamma distribution with logarithm transformation of the mean being modeled as a linear function of predictors • A constant shape parameter for all stations	• Using a gamma distribution • Seperate parameters for each station and each weather state
Spatial dependence	None	Constant inter-site correlation structure	None

SDSM

SDSM is described as a hybrid between a multivariate linear regression method and a stochastic weather generator. Large-scale predictors are used to linearly condition local-scale weather generator parameters (e.g. precipitation occurrence and intensity) at individual stations. Precipitation is then modelled through a stochastic weather generator conditioned on the predictor variables. The conditional probability of precipitation occurrence on day t (ω_t) depends on the large-scale predictors and conditional probability of the previous day's precipitation occurrence (ω_{t-1}) (Wilby et al., 2003). Precipitation occurs if $\omega_t > r_t$ ($0 \leq r_t \leq 1$) where r_t is a uniformly distributed random number. The precipitation amount is modelled through an empirical distribution conditioned on the predictors. The SDSM version 4.2 is used in this study. For a full description, see Wilby et al. (2003).

The SDSM was applied with a common set of predictors to all the stations instead of different sets of predictors for different stations in order to be consistent with the two multi-site models (NHMM and GLIMCLIM). Precipitation was modelled as a conditional process in which local precipitation amounts are correlated with the occurrence of wet days, which in turn correlated with large-scale atmospheric predictors. A transformation of the fourth root was applied to account for the skewed nature of the precipitation distribution. In SDSM, it is possible to adjust the bias correction and variance inflation parameters to overcome the problem of over- or underestimation of the mean and variance of downscaled variables. However, von Storch (1999) indicated that the variance inflation approach adopted in SDSM is not meaningful because it fails to acknowledge that local-scale variation is not completely explained by predictors. In this study, SDSM was run with different adjustments made to the bias correction and variance inflation to test the effect of altering these parameters on downscaled precipitation. However, the alternation of the bias correction and variance inflation did not significantly improve the downscaled results. Therefore, it was decided to use the default value for the variance inflation and no bias correction.

GLIMCLIM

Similar to SDSM, GLIMCLIM also employs a regression-based approach for specifying weather generator parameters conditioned on the large-scale predictors. But, it uses logistic regression to model the rainfall probability and a gamma distribution to model rainfall amounts. The logarithm of the mean precipitation amount is modelled as a linear function of a set of predictors. The shape parameter of the gamma distribution is assumed constant for all observations. The reader is referred to Chandler and Wheater (2002) and Yang et al. (2005) for further details.

In GLIMCLIM, precipitation occurrence and amounts can be modelled with different set of predictors. However, in order to be consistent with SDSM and NHMM a same set of predictors was used to fit the occurrence and amounts model in GLIMCLIM individually. Besides large-scale predictors, other covariates representing spatial dependence, seasonality, autocorrelation and interaction terms can also be used. In this study, site altitudes and the Legendre polynomial transformation of the site eastings and northings are used to accommodate the non-homogeneity displayed across a region that are not explained by the input predictors. The seasonality is represented using sine and cosine components, and the autocorrelation is modelled using the three previous days' rainfall. By defining suitable dependence structures between sites, GLIMCLIM can downscale precipitation at multiple stations simultaneously. For the occurrence model, we used a beta-binomial distribution (see Yang et al. (2005) for a mathematical derivation). The distribution has two parameters representing the mean, which varies in time and is estimated from the probabilities derived from the occurrence model, and the shape, which is assumed constant for all days and is estimated using the method of moments (Chandler and Wheater, 2002; Yang et al., 2005). A small value of the shape parameter indicates strong inter-site dependence. For the amounts model, we used the inter-site constant correlation structure of the transformed rainfall values called the Anscombe residuals (Yang et al., 2005).

NHMM

Unlike SDSM and GLIMCLIM in which the weather generator parameters are conditional directly on the predictors, NHMM conditions the weather generator parameters on weather states. As a weather state-based downscaling model, the NHMM relates synoptic-scale atmospheric predictors through a finite number of 'hidden' (i.e. unobserved) weather states to multi-site daily precipitation occurrences. The temporal evolution of these daily states is modelled as a first-order Markov process with state to state transition probabilities conditional on a set of synoptic-scale atmospheric predictors. The most likely weather state sequence is obtained from a fitted NHMM using the Viterbi algorithm to assign each day to its most probable state (Forney, 1978). Unlike other weather state (type) based downscaling models, the weather states in the NHMM are defined from daily rainfall observations at a network of sites rather than a priori (Hughes et al. 1999).

The NHMM makes two assumptions (Hughes et al., 1999) as shown in Figure 5.1. The first assumption states that the n site precipitation process pattern on day t (R_t) depends only on the weather state of day t (S_t), and the second assumption states that the weather state on day t (S_t) depends both on the weather state on the previous day (S_{t-1}) and the values of the atmospheric variables (X_t) on day t. The two assumptions determine the temporal structure in the precipitation process. The first assumption states that the precipitation process (R_t) is conditionally independent given the weather state. In other words, all the temporal persistence in the precipitation processes is captured by the persistence in the weather state.

Daily precipitation amount at each station is modelled as a combination of a delta function (dry days modelling) and a gamma function (wet days modellling). In this study, the

NHMM computations are performed using the Multivariate Nonhomogeneous Hidden Markov Model toolbox (Kirshner, 2005b). Further details on NHMM can be found in Hughes and Guttorp (1994), Hughes et al. (1999) and Kirshner (2005a).

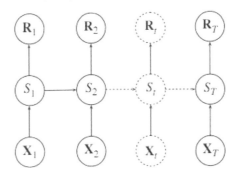

Figure 5.1: The MVNHMM model structure (Source: Kirshner et al., 2005b).

For the NHMM model, calibration is to choose the appropriate number of hidden weather states using the Bayesian Information Criterion (BIC). This involves the sequential fitting of several NHMMs with an increasing number of weather states until the BIC reaches its minimum value. For this study, the appropriate number of hidden weather states is four.

5.2.5. Performance criteria

The standard split-sampling technique of model calibration and validation was implemented in this work. The model calibration was performed for the monsoon seasons (June–September) over the period 1961–1980, while the period 1981–1990 was used for validation. Both the GLIMCLIM and NHMM models are calibrated for the 14-station network concurrently as opposed to the SDSM model, which is calibrated on a station by station basis. The performance of the three downscaling models is evaluated by several criteria relevant to hydrological studies: (1) the spatial correlation structure in terms of Spearman cross correlation in the daily rainfall amounts, (2) the ability to reproduce inter-annual variability in terms of Spearman rank correlation, (3) the mean difference between the observed and simulated data, and (4) the temporal structure (characterized by wet- and dry spell length).

5.3. Results and discussion

The results presented in the following subsection are based on 100 realizations of downscaled precipitation from each of the methods. We also tested the sensitivity of using more number of realizations (e.g. 200, 300, 400 and 500) but found no significant changes in the results.

5.3.1. Validation of the three statistical downscaling models (1981–1990)

Spatial patterns

Accurate reproduction of the spatial pattern of the precipitation is essential for correct simulation of river discharge over a large area. Figure 5.2 presents observed and modelled Spearman cross-correlation for daily precipitation amounts. As can be seen from the figure, NHMM performed quite well in reproducing the spatial correlation for the majority of station pairs. However, it underestimated the spatial correlations for highly correlated stations. This indicates that although the hypothesis of conditional spatial independence, given the weather state, captures much of the correlation between stations, it is not sufficient to account for all the observed correlations between stations. The unexplained local spatial correlation, which is induced by important subgrid-scale features such as topography and convection, was not captured by this assumption. Similar findings were reported by Bellone et al. (2000), Kioutsioukis et al. (2008) and Liu et al. (2012). Both the SDSM and GLIMCLIM models show consistent underestimation of the spatial correlations for most station pairs (Figure 5.3). The possible explanation for the poor performance of SDSM and GLIMCLIM in representing the spatial dependence is that the SDSM as a single-site model is trained on each station separately and therefore could not effectively reproduce the inter-station correlations. GLIMCLIM, which was originally designed for application over smaller areas with frontal (relatively homogeneous) weather systems, models the spatial dependence by constraining it to be the same for all site pairs involved. This is unrealistic in practice, particularly for large areas where inter-site dependence generally tends to be lower and vary with distance. Similar results were obtained in other studies. Yang et al. (2005) discussed the difficulties of GLIMCLIM in representing the spatial dependence over large areas. Frost et al. (2011) found that GLIMCLIM tends to underestimate the spatial correlation with distance under Australian condition. Liu et al. (2012) reported that GLIMCLIM markedly overestimated the spatial correlation at longer distance under north China plain condition. The use of distance-dependent correlation structure in GLIMCLIM is worth investigating in the future.

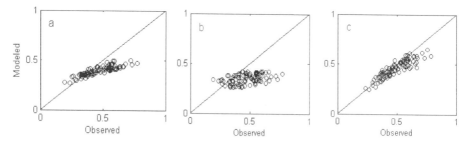

Figure 5.2: Scatter plots of observed and mean modeled Spearman cross correlation obtained by (a) SDSM, (b) GLIMCLIM, and (c) NHMM for the validation period.

Wet and dry spells

Figures 5.3 and 5.4 compare the observed and modelled wet-and dry spell length distributions at three representative stations for the validation period. The results are in general similar for the remaining stations. Overall, both SDSM and GLIMCLIM reproduce wet spell length distribution reasonably well, particularly for the short-duration spells less than 10 days, while it is found that wet spell distribution at some stations is modelled less accurately by NHMM in comparison to other two models. Clearly, this lower performance of NHMM in representing wet-spell distributions may be attributed to its assumption of conditional temporal independence of the precipitation process as discussed in the description of the model in Section 5.2.4. The difficulty in reproducing the wet spell distribution by NHMM was also noted by Hughes and Guttorp (1994) and was attributed to the assumption of conditional temporal independence of the precipitation process in the NHMM. Liu et al. (2012) reported that NHMM performed relatively poorer in reproducing the mean wet- and dry spell length in comparison to GLIMCLIM. As SDSM and GLIMCLIM model the temporal dependence of the precipitation processes by assuming it to be conditionally Markov, i.e. the precipitation process is conditional on the current day's atmospheric variables and the preceding days' precipitation process (the previous day for SDSM and the three previous days for GLIMCLIM), they are able to reproduce wet spell length distribution reasonably well. In comparison to wet spell, all the models show less skill in reproducing dry spell length duration. It should, however, be noted that these are not very frequent events in the monsoon season.

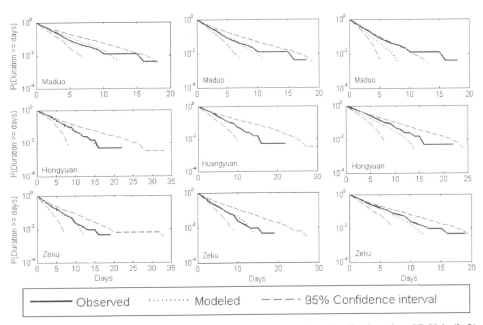

Figure 5.3: Observed versus modelled wet spell lengths distribution by SDSM (left), GLIMCLIM (middle) and NHMM (right) for the validation period at representative stations. Station names refer to Table 2.2 and Figure 2.1.

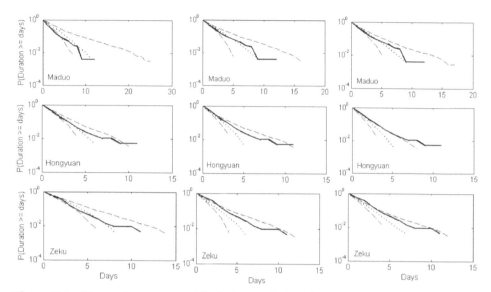

Figure 5.4: Observed versus modelled dry spell lengths distribution by SDSM (left), GLIMCLIM (middle) and NHMM (right) for the validation period at representative stations. Station names refer to Table 2.2 and Figure 2.1.

Inter-annual variability and the magnitude of observed summer precipitation

It is important for the downscaling models to be able to reproduce the inter-annual variability reasonably well if they are to be used in climate change studies; otherwise, the models lack in sensitivity to climate variability, and their usefulness in climate change studies can be questioned (Wetterhall et al. 2006). To estimate the inter-annual variability, we calculate the Spearman rank correlation between the observed and median of 100 simulations at each station for the validation period (Figure 5.5). The horizontal line in this plot shows the value of the correlation coefficient above which they are statistically significant at the 95 % confidence level. Overall, NHMM performs relatively better than the other two models in reproducing the inter-annual monsoon precipitation variability. This can be seen from the plot that NHMM exhibits relatively high correlation coefficients which are statistically significant at the majority of the stations (9 out of 14 stations). However, GLIMCLIM and SDSM perform less satisfactorily in reproducing interannual monsoon precipitation variability with most stations (8 and 10 out of 14 stations, respectively) having low and insignificant correlation coefficients. This could be due to the reason that the two regression-based models are not able to capture some processes (e.g. localized convection) that are driving inter-annual precipitation variability because only part of the local climate variability is related to large-scale climate variations. The introduction of parametric inflation factor in the SDSM is found to be ineffective to sufficiently represent variability in the downscaled precipitation (see von Storch (1999) on the limitation of using inflation in downscaling to increase variance). However, through a number of weather states defined from the 14-station rainfall observations, NHMM is able to capture local precipitation variability reasonably well.

Figure 5.6 presents the percentage difference between the observed and median of 100 simulations at each station for each validation year. The whisker–box plots show the biases across all stations. We can see that NHMM captures the magnitude of the observed monsoon precipitation sufficiently well with much lower biases for almost the whole validation period.

However, the other two models perform less satisfactorily with an overestimation at most stations for most of the validation period.

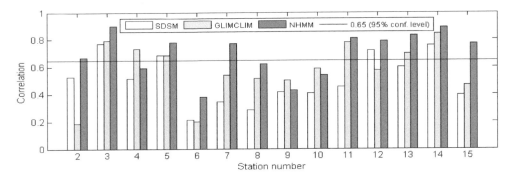

Figure 5.5: Correlation between the simulated and observed summer precipitation for each station during the validation period 1981–1990. Station numbers refer to Table 2.2 and Figure 2.1.

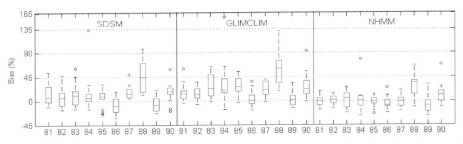

Figure 5.6: Box-plots of bias (percentage difference between observed and median simulated) in downscaled summer precipitation during the validation period 1981-1990. The box-plots depict the range of the bias across 14 stations. The boxes denote the median and interquartile range (IQR). Whiskers extend 1.5 IQR from box ends, with outliers denoted as "o".

5.3.2. Downscaling precipitation for the present climate (1960–1990)

The downscaling models calibrated and validated using the NCEP predictors were driven by the two GCM predictors for the present climate (1961–1990) to evaluate whether downscaled summer precipitation from the two GCMs can reproduce the variability of the observed one. Figure 5.7 presents the percentage difference between downscaled and observed summer precipitation. The results show that the biases in the downscaled summer precipitation are quite similar from GCM to GCM, while they vary considerably from downscaling model to downscaling model. The NHMM appears to be the best performer when driven by both the CGCM3 and the ECHAM5 predictors with the biases ranging from −2.5 to 1.3 % across different stations. SDSM generally shows large positive bias in the downscaled summer precipitation compared to those from the NHMM. The GLIMCLIM shows mostly negative biases (underestimation), which are relatively larger (−5 to−20 %) than those of the other two models.

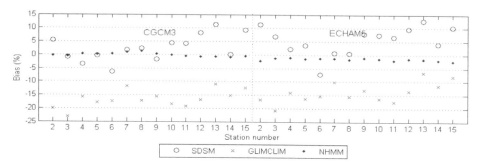

Figure 5.7: Bias (percentage difference between observed and median simulated) in downscaled summer precipitation from the CGCM3 and ECHAM5 predictors at each station. Station numbers refer to Table 2.2 and Figure 2.1.

5.3.3. Downscaling precipitation for the future scenarios (2046–2065)

Three statistical models (calibrated) are used to downscale daily precipitation from two GCMs for three emission scenarios. Estimated changes in the magnitude and the distribution of six precipitation indices for a future period (2046–2065) are investigated against the control period (1961– 1990). The changes in the magnitude correspond to the percentage difference between mean values of each index in the future period and those in the control period. A two-tailed Student's t test for the 5 % confidence level is performed to check if the mean values from the present and future periods are significantly different.

Estimated changes in the magnitude of precipitation indices

Figure 5.8 depicts the changes in the magnitude of the six precipitation indices. We can see that there is strong consistency in the climate change signals from different projections. All of the projections suggest an increase in the indices related to the wet events (*prcptot, pq95, pq95tot, pfl95* and *px5d*) and a decrease in the index related to the dry events (*pxcdd*). The effect of the driving GCM on the magnitude of estimated changes is evident in Figure 5.8, with CGCM3-driven projections showing relatively larger changes in the precipitation indices than ECHAM5-driven ones. A comparison between the three downscaling models shows that for all the indices considered, SDSM predicts larger changes than GLIMCLIM and NHMM. However, despite some notable differences in the results for the control climate, it is interesting to note that the projected changes by GLIMCLIM and NHMM are of similar magnitude. Compared to the differences due to the GCMs and downscaling models, there are no clear systematic differences between the projected changes for the three emission scenarios. The SDSM also shows large spatial variability of the projected changes across stations, while the other two models show less spatial variability, especially for the four wet extreme indices (*pq95, pq95tot, pfl95* and *px5d*). This is probably due to the fact that SDSM is calibrated on individual station, while other two models are calibrated on multi-station basis.

Throughout the study region, all of the projections show statistically significant increases in summer precipitation (*prcptot*) ranging from an average of 8 to 55 %. Similar results are obtained for the *pq95tot* index, but with much larger magnitude. On average, *pq95tot* are expected to increase by 13 to 167 %. As for the *pq95* and the *pfl95* indices, the majority of the projections suggest a statistically significant increase at almost all stations with the exception of the ECHAM5/GLIMCLIM-driven one in which most stations reveal insignificant increases. Similar to the *prcptot* and the *pq95tot* indices, there is also a pronounced increase in the consecutive 5-day precipitation total (*px5d*) over the whole region with an average of 5 to 60 %. In clear contrast to all the wet indices, the maximum dry spell (*pxcdd*) is expected to decrease over the whole region, but most of the decreases are statistically insignificant.

5.3.4. Changes in the distribution of precipitation indices

In this section, we analysed changes in the distribution of the precipitation indices. Figure 5.9 compares fitted gamma probability density functions (PDFs) of the precipitation indices averaged across 14 stations for the control and future periods. Comparison of the scenario and control PDFs reveals a substantial shift of the mean to the right of the distribution for the wet indices and a small shift to the left for the dry index, suggesting a pronounced increase in the precipitation indices related to the wet events and a small decrease in the maximum dry spell. Concerning the shape of the distribution, it is noteworthy that the scenario PDFs of the wet indices generally become wider and flatter in comparison to that from the control period, which may suggest increased variability in future precipitation.

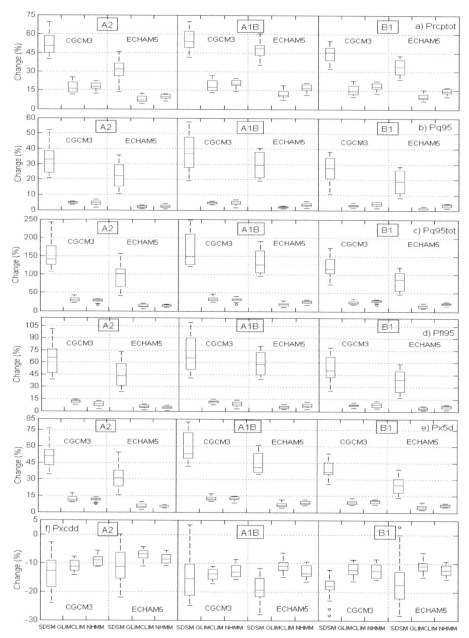

Figure 5.8: Box plots of projected precipitation indices anomalies (A2, A1B and B1 scenarios, 2046-2065 minus 1961-1990) based on downscaled results from CGCM3 and ECHAM5. The box-plots depict the range of projected precipitation anomalies across 14 stations. The boxes denote the median and interquartile range (IQR)). Whiskers extend 1.5 IQR from box ends. Spatial variability can be inferred from the height of the box and whiskers.

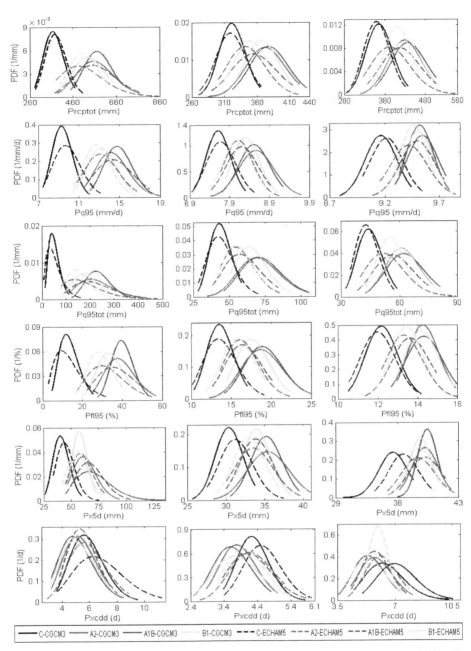

Figure 5.9: Fitted gamma probability density functions (PDF) for future (2046-2065) and current (1961-1990) precipitation indices averaged across all stations based on downscaled results from two GCMs. SDSM (left column); GLIMCLIM (middle column); NHMM (right column). C denotes the control climate; A2, A1B and B1 denote the three emission scenarios.

The climate change signal derived from the present study implies a significant increase in summer precipitation totals and extremes, and an insignificant decrease in the maximum dry spell. This signal is in general agreement with previous modelling studies over the neighbouring areas, which may be interpreted as the warmer air in the future climate being able to hold more moisture generated by increased evaporation from warmer oceans. When this moister air moves over land, more intense precipitation is produced (Meehl et al., 2005). Similar projections for 2081–2100 under the A1B scenario were obtained for annual precipitation extremes over the headwater region of the Yangtze River by Xu et al. (2009a) based on statistical downscaling of six GCM outputs using the SDSM model. Increases in extreme precipitation have also been found from the direct GCM outputs over a large scale (Xu et al., 2011; Yang et al., 2012). Based on direct outputs from three GCMs, Xu et al. (2011) suggested increases in the px5d and pfl95 indices and little change in the pxcdd during summer over the Huang-Huai-Hai River Basins for 2011–2050 under the A1B scenario. Using the ensemble mean of five GCMs, Yang et al. (2012) reported slight decreases in the pxcdd and general increases in the px5d and pfl95 indices over most of the Tibetan Plateau by the end of the century under the three scenarios (A2, A1B and B1).

5.4. Conclusions

Three statistical downscaling models have been compared in terms of their ability to downscale summer (June–September) daily precipitation over the source region of the Yellow River. These models were then applied to investigate possible changes in rainfall totals and extremes by the middle of the twenty-first century using the predictors from two GCMs (CGCM3 and ECHAM5) under the IPCC SRES A2, A1B and B1 scenarios. The validation (1981–1990) results show that the NHMM model is generally better in reproducing the spatial correlation structure, inter-annual variability and magnitude of observed summer precipitation in comparison to other two models. The NHMM, however, has difficulty in reproducing the observed wet and dry spell length distribution. In contrast, SDSM and GLIMCLIM show consistent underestimation of the spatial correlations for most station pairs. This is due to the fact that the single-site model SDSM was trained on each station separately, and the multi-site model GLIMCLIM simulates the spatial rainfall dependence structure by constraining it to be the same for all site pairs involved, which makes the model less capable of reproducing the spatial correlation structure over such large study area. Conditional on current day's atmospheric variables and precipitation process of preceding days, temporal dependence at short durations is generally preserved well by SDSM and GLIMCLIM, while it is modelled less satisfactorily by NHMM at some stations. The better reproduction of local precipitation variability by NHMM may be attributed to the fact that it makes use of a number of weather states defined from the 14-station rainfall observations.

For future projection, there is a strong consistency in the climate change signal derived from the application of three statistical models to downscale precipitation from two GCMs and three emission scenarios. Overall, all parts of the study region is expected to experience a significant increase in rainfall totals and extremes, accompanied by an insignificant reduction in the maximum dry spell. The climate change signal presented here is physically consistent with warmer air in the future climate being able to hold more moisture generated by increased evaporation from warmer oceans. Although there is strong agreement in the direction of the projected changes, there is large uncertainty in the magnitude of the changes. A large amount of uncertainty is found to be associated with the choice of a downscaling method. In addition, for most indices the scenario PDFs show large shift and become flatter compared to the control period, suggesting that the increase in the magnitude of rainfall totals and extremes is accompanied by an increase in their inter-annual variability.

Overall, this charpter highlights the importance of acknowledging limitations and advantages of different statistical downscaling methods, and it also implies that climate projection based on only one GCM, one downscaling model or one emission scenario have to be interpreted with caution.

6. Expected changes in future temperature extremes and their elevation dependency over the YRSR[5]

Abstract: Using the Statistical DownScalingModel (SDSM) and the outputs from two global climate models, we investigate possible changes in mean and extreme temperature indices and their elevation dependency over the Yellow River source region for the two future periods 2046–2065 and 2081–2100 under the IPCC SRES A2, A1B and B1 emission scenarios. Changes in interannual variability of mean and extreme temperature indices are also analyzed. The validation results show that SDSM performs better in reproducing the maximum temperature-related indices than the minimum temperature-related indices. The projections show that by the middle and end of the 21st century all parts of the study region may experience increases in both mean and extreme temperature in all seasons, along with an increase in the frequency of hot days and warm nights and with a decrease in frost days. By the end of the 21st century, interannual variability increases in the frequency of hot days and warm nights in all seasons and frost days in spring while it decreases in frost days in summer. For autumn pronounced elevation-dependent changes are observed in which around six out of eight indices show significant increasing changes with elevation.

6.1. Introduction

The YRSR is situated in the northeast Tibetan Plateau, which has been identified as a "climate change hot-spot" and one of the most sensitive areas to greenhouse gas (GHG)-induced global warming (Giorgi, 2006). This region is geographically unique, possesses highly variable climate and topography, and plays a critical role for downstream water supply. A growing number of evidences suggest that this region and its surroundings are experiencing warming and accelerated glacier retreat (Liu and Chen, 2000; You et al., 2008; Liu et al., 2009; Qin et al., 2009; Rangwala et al., 2009; Hu et al., 2011, 2012; Immerzeel et al., 2010; Maskey et al., 2011; Shrestha and Aryal, 2011). In line with global climate projection, this warming is expected to continue into the future under enhanced greenhouse gas forcing (IPCC, 2007).

A primary concern in estimating impacts from climate changes are the potential changes in variability and hence extreme events that could be associated with global climate change (Marengo et al., 2010). Recent model studies (based on both global and regional climate models) suggest that the 21st century is very likely to be characterized by more frequent and intense temperature extremes, which are not only due to the mean warming, but also due to changes in temperature variability (IPCC, 2007; Tebaldi et al., 2006; Kjellström et al., 2007; Fischer and Schär, 2009). Regional temperature extremes have recently received increasing attention given the vulnerability of our societies to such events. This is particularly true for mountain regions where the observed or projected warmings are generally greater than at low-elevation regions (Diaz and Bradley, 1997; Beniston et al., 1997; Rangwala et al., 2009; Liu et al., 2009; Qin et al., 2009; Rangwala and Miller, 2012; Viviroli et al., 2011). Moreover, some mountain regions have demonstrated an elevation dependency in surface warming (i.e. greater warming rates at higher altitude) in the latter half of the 20th century and/or during the 21st century (Beniston and Rebetez, 1996; Diaz and Bradley, 1997; Giorgi

―――――――――――――
[5] This chapter is based on paper Expected changes in future temperature extremes and their elevation dependency over the Yellow River source region, China by Hu, Y., Maskey, S. and Uhlenbrook, S. 2013. Hydrology and Earth System Science 17: 2501-2514. DOI: 10.5194/hess-17-2501-2013.

et al., 1997; Liu and Chen, 2000; Chen et al., 2003; Diaz and Eischeid, 2007; Rangwala et al., 2009, 2010; Liu et al., 2009).

Within the Tibetan Plateau, previous studies found indications for enhanced warming at higher elevation (Liu and Chen, 2000; Chen et al., 2003; Liu et al., 2009; Qin et al., 2009; Rangwala et al., 2009, 2010), while others reported no enhanced or even weakening warming at higher elevations (You et al., 2008; Lu et al., 2010). Although a number of climate change studies over the YRSR have been reported in the literature, possible changes in future temperature extremes and their relationship with elevation are yet to be fully explored. Earlier studies Xu et al. (2009) and Wang et al. (2012) reported increases in the mean (Tmean), maximum (Tmax), and minimum (Tmin) temperature over this region for the 21st century.

This study complements the previous studies by including estimated changes in future temperature extremes using a number of indices and their elevation dependency. Changes in interannual temperature variability are also examined in the present study. Among different downscaling approaches, statistical downscaling is the most widely used one to construct climate change information at a station or local scales because of its relative simplicity and less intensive computation. Moreover, previous studies reported in the literature found statistical downscaling showing similar skill as dynamical downscaling and no indication of either downscaling method having a direct advantage over the other (Haylock et al., 2006; Schoof et al., 2009). In the present study the Statistical DownScaling Model (Wilby et al., 2002) is applied to downscale the outputs of the two driving GCMs under the IPCC SRES A2, A1B and B1 emission scenarios.

6.2. Material and methods

6.2.1. Data sets

Observed station data

Daily maximum and minimum temperature from 13 stations sparsely distributed throughout the study region, for the period 1961–1990 were used in this study. Figure 2.1 depicts the geographical location of the stations in the study region and Table 2.2 shows their latitude, longitude and altitude. Slightly less than 0.0017% of the data from two stations were missing, which were infilled using the records from neighboring stations. The double mass curve method was applied to test the homogeneity of the data set by plotting the monthly value from the station against the mean values (monthly) of all other stations (Hu et al., 2012). According to the results of the test, all the data were found homogeneous.

Reanalysis data

In addition to the observed data, large-scale atmospheric predictors derived from the National Center for Environmental Prediction/National Centre for Atmospheric Research (NCEP/NCAR) reanalysis data set (Kalnay et al., 1996) on a $2.5° \times 2.5°$ grid over the same time period as the observation data were employed for calibration and validation of the statistical downscaling models. These variables include specific humidity, air temperature, zonal and meridional wind speeds at various pressure levels and mean sea level pressure.

GCM data

In order to project future scenarios, outputs from two GCMs under the Intergovernmental Panel on Climate Change Special Report on Emissions Scenarios (IPCC-SRES) A2, A1B and B1 were used. For details on the two GCMs and the three emission scenarios, please refer to section 5.2.1. Those two GCMs were selected on the basis of (i) their relatively reasonable performances in simulating the 20th century surface air temperature over China (Zhou and

Yu, 2006; Wang et al., 2013) and the South Asian summer monsoon over the historical period (Fan et al., 2010) and (ii) their wide use in previously conducted climate change studies. The GCM simulations corresponding to the present (1961–1990) and two future climates (2046–2065 and 2081–2100) were considered in the analysis. Prior to use in this study, both GCMs grids were linearly interpolated to the same 2.5° ×2.5° grids fitting the NCEP reanalysis data.

6.2.2. Temperature indices

To represent extreme temperature conditions (both the frequency and intensity of temperature extremes), eight temperature indices, including two indices for mean minimum and maximum temperature, are selected. The indices included in this study are:

1. mean Tmax (Txav) – mean daily maximum temperature [∘C];
2. mean Tmin (Tnav) – mean daily minimum temperature [∘C];
3. diurnal temperature range (DTR) – difference between daily maximum and minimum temperature [∘C];
4. hot day (Txq90) – 90th percentile value of daily maximum temperature in a year [∘C];
5. cold day (Tnq10) – 10th percentile value of daily minimum temperature in a year [∘C];
6. frequency of hot days (Tx90p) – the percentage of time in a year when daily maximum temperature is above the 90th percentile of the 1961–1990 daily maximum temperature distribution [%];
7. frequency of warm nights (Tn90p) – the percentage of time in a year when daily minimum temperature is above the 90th percentile of the 1961–1990 daily minimum temperature distribution [%]; and
8. frost days (Tnfd) – the number of days with daily minimum temperature <0 ∘C [days].

Each of the indices has been calculated for 1961–1990 (present period) and 2081–2100 (future period), and for three scenarios A2, A1B and B1. Except for the frost days, all of the indices have been analyzed for four seasons, which are defined as winter (December–February, DJF), spring (March–May, MAM), summer (June–August, JJA) and autumn (September–November, SON). Frost days are not analyzed for winter since it has little meaning for the study region in winter where the daily minimum temperature is around 20 ∘C below zero.

6.2.3. Choice of predictors

Following the steps described in section 5.2.3, a number of predictors for downscaling Tmax and Tmin was selected as follows: specific humidity at 700, 850 and 1000 hPa level and air temperature at 500, 700, 850 and 1000 hPa level. The same predictor domain as used in downscaling precipitation was used for downscaling temperature here. The predictors were first standardized at each grid point by subtracting the mean and dividing by the standard deviation. A principal component analysis (PCA) was then performed to reduce the dimensionality of the predictors. The first eight principal components (PCs), which account for more than 90% of the total variance, were then used as input to the downscaling model.

6.2.4. Statistical downscaling model (SDSM)

The SDSM were selected to downscale temperature. For temperature the downscaled process is unconditional,i.e. there is a direct linear relationship between the predictand (i.e. temperature) and the chosen predictors. For more information on the SDSM, refer to section 5.2.4.

6.3. Results and discussion

6.3.1. Validation of the statistical downscaling model (validation period 1981–1990)

The standard split-sampling technique of model calibration and validation was implemented in this work. The model calibration was performed for the period 1961–1980, while the period 1981–1990 was used for validation. As the SDSM is a stochastic model, 100 realizations of daily maximum (minimum) temperature are generated, and the indices are calculated as the average of the indices calculated from each realization. We also tested the sensitivity by using a larger number of realizations (e.g. 200, 300, 400 and 500) but found no significant changes in the results. The skill of the downscaling model to reproduce the mean and extreme temperature is evaluated and compared in terms of the Spearman's rank correlation coefficient and the bias between the simulated and observed indices. Model evaluation was performed on a monthly basis.

Figure 6.1 shows the correlation coefficients and the differences between the simulated and observed indices (mean maximum and minimum temperature, 90th percentile of the maximum temperature, and 10th percentile of minimum temperature) for each month. The whisker-box plots show spatial variability of the correlations and the bias across all the stations. The horizontal solid line in Fig. 6.1a–b shows the value of the correlation coefficient above which they are statistically significant (95%confidence level). As can be seen from these plots, the model simulates the mean and the 90th percentile of daily maximum temperature (Txav, Txq90) very well with the majority of the stations showing statistically significant correlations and relatively lower biases in almost all months. However, a relatively poor performance in simulating Txav in August and October and Txq90 in September and December is observed. The mean minimum temperature (Tnav) is also well reproduced by the models in most months with the exception of the winter months from November to January. The model performance is generally poor for the 10th percentile of daily minimum temperature (Tnq10) where most stations show insignificant correlations and large bias in most months. Generally, the model shows more skill for the maximum temperature-related indices (Txav, Txq90) than for the minimum temperature-related indices (Tnav, Tnq10). A comparison between different months reveals that in general the temperature indices were better downscaled for the summer months than for other months. Such a seasonal dependence of downscaling skill was also found in other parts of the world (e.g. Haylock et al., 2006 in England; Wetterhall et al., 2007 in Sweden; Hundecha and Bárdossy, 2008 in German). This may relate to the fact that the local climate of the study region in summer is largely determined by large-scale circulation (e.g. summer monsoon) while it is mainly determined by local convective processes in other seasons.

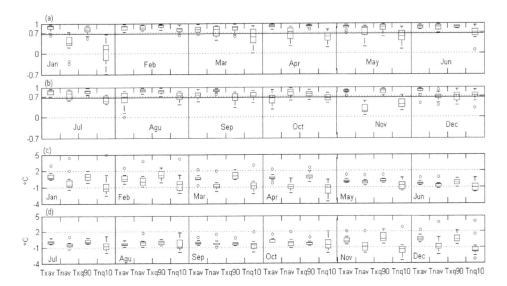

Figure 6.1: Correlations (a–b) and differences (c–d) between the simulated and the observed extreme temperature indices for each month during the validation period 1981–1990. The whisker-box plots depict the range of the correlation across 13 stations. The boxes denote the median and interquartile range (IQR). Whiskers extend 1.5 IQR from box ends, with outliers denoted as circles. The horizontal solid line denote significant correlation at the 5% confidence level.

6.3.2. Downscaling for the current climate (1961–1990)

The downscaling model calibrated and validated using the NCEP predictors was forced by the two GCMs outputs for the present climate (1961–1990) to evaluate whether the downscaled temperature indices from the two GCMs can reproduce the variability of the observed ones. Figure 6.2 depicts the difference between the downscaled and observed temperature indices (Txav, Tnav, Txq90, Tnq10) for each station and each season. Overall, the downscaled results from both GCMs are able to reproduce the observed temperature indices reasonably well with the bias generally varying between −2 and 2 °C across different stations in all seasons with a few exceptions in winter and autumn. The biases in the downscaled temperature indices are of similar magnitude for the two GCMs, and no systematic and notable differences are found.

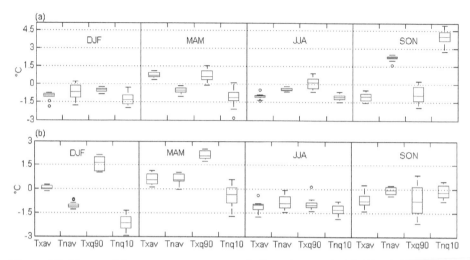

Figure 6.2: Biases of the extreme temperature indices downscaled from the CGCM3 (a) and the ECHAM5 predictors (b) for the four seasons during the control period 1961-1990. The whisker-box plots depict the range of the bias across 13 stations. The boxes denote the median and interquartile range (IQR)). Whiskers extend 1.5 IQR from box ends, with outliers denoted as circles.

6.3.3. Future projections (2046–2065 and 2081–2100)

The statistical downscaling model (calibrated) is used to downscale daily maximum and minimum temperature from two GCMs for three emission scenarios. Estimated changes in the mean of selected temperature indices for the two future periods (2046–2065 and 2081–2100) are investigated against the control period (1961–1990). The changes in the mean correspond to the difference between mean values of each index in the future period and those in the control period. A two-tailed Student's t test for the 5% confidence level is performed to check if the mean values from the present and future periods are significantly different. Also, we have analyzed elevation dependency of the projected changes for each index. Figures 6.3–6.6 illustrate the projected climate change of each index with station altitude. A one-tailed Student's t test for the 5% confidence level is performed to check if the linear trends of the projected changes with increasing altitude are statistically significant.

Projected changes in the mean state of the temperature indices

All the temperature indices, with the exception of the DTR, show statistically significant warming at all stations in all future seasons with both GCMs and three emission scenarios (Figures 6.3–6.6). By the middle and end of the 21st century, all parts of the study region are expected to experience statistically significant increases in the intensity of both mean and extreme temperature, together with significant increases in the occurrence of hot days and warm nights and with decreases in frost days. As expected, the projected changes in the temperature indices for 2081–2100 are generally larger than those for 2046–2065. The accelerated warming suggested by these results may be due to the strong greenhouse forcing toward the end of the 21st century. While there is strong agreement in the direction of projected changes, the magnitude of the changes varies between different GCMs and emission scenarios. The effect of the driving GCM on the magnitude of estimated changes in 2081–2100 is evident in Figs. 6 and 7, with the ECHAM5-driven projections showing larger changes in the temperature indices than the CGCM3-driven ones. Also, we note that the projected changes in 2081–2100 tend to scale with the emission scenario, i.e. the larger the greenhouse gas forcing, the stronger the response (generally most intense in the A2, followed by the A1B and B1 scenarios). However, the same does not hold true in 2046–2065, where no systematic differences in the magnitude of the projected changes from two GCMs are noticed (Figures 6.3 and 6.4). Unlike the end of the 21st century, we note that in some cases the projected changes for the middle of the 21st century are stronger in the A1B scenario than in the A2 scenario. This is particularly noticeable in the ECHAM5-driven projections. This is probably due to the following reason: although the CO_2 concentrations are similar in A1B and A2 emission scenarios up to the middle of the 21st century, the A2 scenario specifies somewhat greater sulphate aerosol concentrations, which are thought to have a cooling effect on surface temperature (Ramanathan et al., 2001).

For both the future periods, we see a similar and pronounced seasonality of projected changes. For the intensity-related indices, the mean maximum temperature and hot day (Txav, Txq90; Figure. 6.3a and b and 6.5a and b) show the largest warming in winter and the least one in summer, while the mean minimum temperature and cold night (Tnav, Tnq10; Figures 6.3c and d and 6.5c and d) show the largest warming in autumn and the least one in spring, which is partly consistent with recent observations over the study region and the Tibetan Plateau, where winter was reported to have the largest warming rate, followed by autumn (Hu et al., 2012; Liu and Chen, 2000; Liu et al., 2006; Rangwala et al., 2009; Zhang et al., 2008). However, note that there are some discrepancies in the seasonality of projected warming as reported in different studies, which is probably due to choice of different GCMs. For example, using the same downscaling model (SDSM) but with a different GCM (HadCM3), Wang et al. (2012) found that Txav and Tnav is expected to undergo the largest warming in autumn and summer, respectively, during the period 2070–2099 under the A2 and B2 scenarios. Using several GCMs (CGCM2, CCSR, CSIRO and HadCM3), Xu et al. (2009) reported that Txav (Tnav) would experience greater warming in spring and autumn (summer and autumn) under the B2 scenario. Compared to other temperature indices, projected changes in diurnal temperature range (DTR) are less strong and less consistent. DTR is expected to experience a significant decrease in summer and autumn, indicating a greater warming in minimum temperature than in maximum temperature, consistent with recent observational studies over this region and its vicinity (Hu et al., 2012a; Liu et al., 2006; You et al., 2008). However, changes in DTR are ambiguous in winter and spring with the CGCM3-driven projections showing non-significant decreases and the ECHAM5-driven ones significant increases. As for the frequency-related indices (Figures 6.4 and 6.6), the occurrences of hot days and warm nights show the largest increases in summer and the least ones in spring, while frost days show the largest decrease in summer and autumn. Under the same emission scenarios, T. Yang et al. (2012) reported similar findings for the frost days and the frequency of warm nights over the entire Tibetan Plateau for the 21st century based on multi-model ensemble projections.

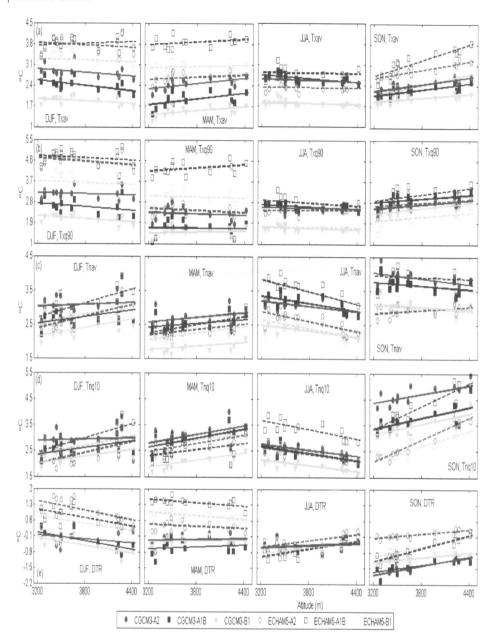

Figure 6.3: Projected anomalies of the intensity-related indices (between 2046-2065 and 1961-1990) with station altitude for four seasons based on statistical downscaling outputs of two GCMs (CGCM3 and ECHAM5) under three emission scenarios (A2, A1B and B1).

Elevation dependency of the projected changes in the temperature indices

As displayed in Figures 6.3a and 6.5a, the projected warming in autumn Txav shows a statistically significant increasing trend with altitude in the two future periods, with a varying rate of 0.2–0.9 °C per km for 2046–2065 and 0.48–1.1 °C per km for 2081–2100, respectively. A similar tendency is found in spring, but only the trends based on the CGCM3 projections are statistically significant. As in the case of Txav, there is also a pronounced elevation dependency in projected warming in autumn Txq90 in the period 2081–2100 but at a lesser rate (0.23–0.73 °C per km) (Fig. 6b). A similar tendency is observed for autumn Txq90 in the period 2046–2065 but insignificant in most cases. For Tnav in summer (Figures 6.3c and 6.5c), a significant decreasing warming with altitude is noted with a rate ranging from 0.26 to 0.61 °C per km for 2046–2065 and 0.28 to 0.69 °C per km for 2081–2100. In contrast, winter and spring demonstrate increasing warming with altitude for the two future periods, but only the trends based on the ECHAM5 projections reach the significance level. The results reported here are in overall agreement with the findings obtained by Xu et al. (2009) over the Tibetan Plateau, suggesting elevation-dependent warming in Tnav in all seasons other than summer for the end of the 21st century under the A1B scenario. Similar to Tnav, future warming in summer Tnq10 suggests a significant decreasing trend with elevation (Figures 6.3d and 6.5d). This is in clear contrast to other seasons, in particular to autumn, where a strong elevation-dependent warming demonstrates with a much larger rate of 0.55–1.48 °C per km for 2046–2065 and 0.72–2.6 °C per km for 2081–2100. Concerning DTR (Figures 6.3e and 6.5e), it is unexpected to see that future reductions in this index during autumn show a significant weakening tendency with altitude. Similar results are projected for summer, but only the trends based on the ECHAM5 projections reach the significance level. Future increases in autumn Tx90p show a significant increasing trend with elevation (Figures 6.4a and 6.6a). A similar trend is also projected for Tx90p in winter with the ECHAM5-driven projections and in spring with the CGCM3-driven projections. Projected increases in spring Tn90p in 2081–2100 show a significant increasing trend with elevation. Similar trends are noted for spring Tn90p in 2046–2065, but only the trends based on the ECHAM5 projections are statistically significant. Regarding future reductions in frost days (Figures 6.4c and 6.6c), note that summer shows a strong enhanced decrease with elevation at a rate of 8–14 days per km for 2046–2065 and 10–23 days per km for 2081–2100 while spring and autumn show an opposite trend at a lesser rate.

In general, the two future periods show similar elevation-dependent changes with the rate of the projected changes, with altitude being stronger in 2081–2100 than in 2046–2065. The indices related to the minimum temperature demonstrate more pronounced elevation-dependent changes than the indices related to the maximum temperature. In comparison to other seasons, autumn shows pronounced elevation-dependent changes in which around six out of eight indices show significant increasing changes with elevation. By investigating trends on the observed data from the latter half of the 20th century over the same region, Hu et al. (2012) also showed more pronounced elevation-dependent changes in the indices related to the minimum temperature. However, in their study winter season indices showed more pronounced elevation-dependent changes than other seasons.

Projected changes in interannual variability of the temperature indices

The analysis of changes in interannual variability of each index has been done by applying an *F* test on the variance of estimated Probability Density Functions (PDFs) of the future and control periods at the 5% level. For the period 2046–2065, the future PDFs of all the indices show insignificant changes in the shape in all seasons in comparison to a substantial shift of the mean (not shown). Similar results were obtained for the intensity-related indices in the period 2081–2100 (not shown). However, the same is not true in the case of the frequency-related indices in 2081–2100, where the future PDFs show a large shift of the mean as well as significant changes in the shape (Figure 6.7). By the end of the 21st century, the future PDFs of the frequency-related indices become wider and flatter in all seasons for the occurrence of hot days and warm nights and for frost days in spring while they become narrower and sharper for frost days in summer. This suggests that by the end of the 21st century the interannual variability of the occurrence of hot days and warm nights might increase in all seasons while that of frost days might decrease in summers and increase in springs.

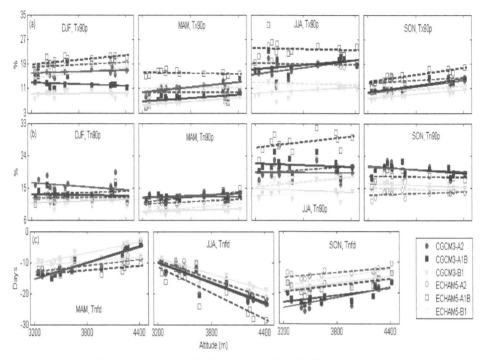

Figure 6.4: As in Figure 6.3, but for the frequency-related indices.

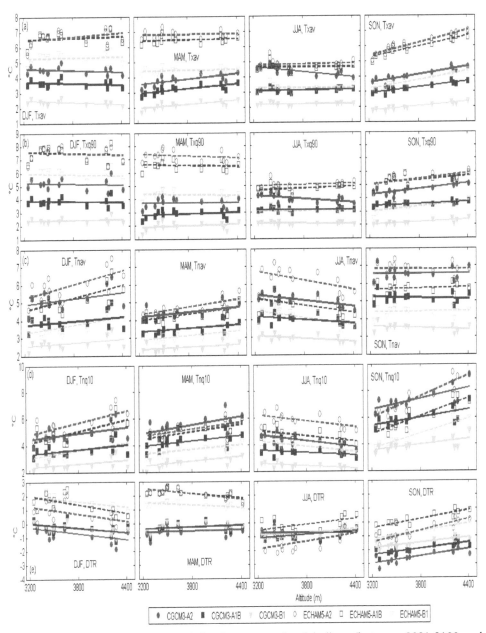

Figure 6.5: Projected anomalies of the frequency-related indices (between 2081-2100 and 1961-1990) with station altitude for four seasons based on statistical downscaling outputs of two GCMs (CGCM3 and ECHAM5) under three emission scenarios (A2, A1B and B1).

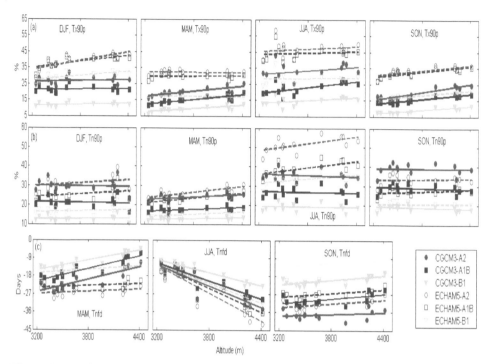

Figure 6.6: As in Fig. 6, but for the frequency-related indices.

6.4. Conclusions

This study presents projections of possible changes in mean and extreme temperature indices and their elevation dependency over the Yellow River source region for the two future periods 2046–2065 and 2081–2100 (relative to 1961–1990) under the SRES A2, A1B and B1 emissions scenarios. The projections are performed using the Statistical DownScaling Models (SDSM) to downscale the outputs of two GCMs (CGCM3 and ECHAM5). Validation results using the NCEP/NCAR reanalysis data show that SDSM performs better in reproducing the maximum temperature-related indices than the minimum temperature-related ones. When driven by the GCMs outputs corresponding to the control period 1961–1990, the downscaled temperature indices are able to reproduce the observed ones reasonably well with the two GCMs showing similar bias.

For the middle and end of the 21st century, all parts of the study region are expected to undergo significant increases in the intensity of mean and extreme temperature in all seasons along with significant increases in the frequency of hot days and warm nights and with decreases in frost days. As expected, the projected changes in the temperature indices in 2081–2100 are generally larger than those in 2046–2065. Compared to other indices, changes in diurnal temperature range are less significant and less consistent in winter and spring. Diurnal temperature range is expected to experience a significant decrease in summer and autumn, indicating a greater warming in minimum temperature than in maximum temperature. The two future periods show similar elevation-dependent changes with the rate of the projected changes with altitude being stronger in 2081–2100 than in 2046–2065. Many of the indices demonstrate elevation-dependent changes, which varies from index to index and from season to season. All the intensity-related indices show a significant increasing warming with elevation in autumn with the exception of Tnav. In contrast, projected warming in Tnav and Tnq10 in summer displays a significant decreasing trend with elevation. Projected increases in hot days and warm nights show a significant increasing trend with elevation in autumn and spring, respectively. A similar trend is also found for the reductions in frost days in summer. However, reductions in frost days tend to decrease with elevation in spring and autumn, with the majority of the projections reaching the significance level. Along with a large shift of the mean, significant changes in the shape of the future PDFs are also observed for the frequency-related indices in 2081–2100, indicating significant changes in interannual variability. By the end of the 21st century, the frequency of hot days and warm nights is likely to experience significant increasing in interannual variability in all seasons under the considered future scenarios. Frost days are expected to experience significant decreasing in interannual variability in summers and increasing one in springs.

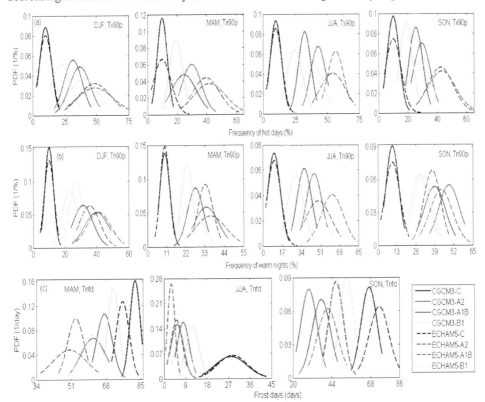

Figure 6.7: Fitted normal probability density functions (PDFs) of the frequency-related indices averaged across stations for 2081-2100 and 1961-1990 for four seasons based on statistical downscaling outputs of two GCMs (CGCM3 and ECHAM5) under three emission scenarios (A2, A1B, B1).

7. Impacts of climate change on the hydrology of the YRSR[6]

Abstract: This chapter investigates the potential impacts of climate change on the hydrology of the Yellow River source region, a large-scale mountainous catchment of critical importance for China with regard to water resources. A fully distributed, physically based hydrologic model (WaSiM) was employed to simulate baseline (1961-1990) and future (2046–2065 and 2081–2100) hydrologic regimes based on climate change scenarios derived from statistically downscaling two global climate models (GCMs) under three emissions scenarios (B1, A1B and A2). All climate change projections show year-round increases in both precipitation and temperature, which result in significant increases in streamflow and evaporation on both annual and seasonal basis. High flow is expected to increase considerably in most projections, whereas low flow is expected to increase slightly. Snow storage is projected to considerably decrease while the peak flow is likely to occur later. We also observe a significant increase in soil moisture on annual basis owing to increased precipitation. Overall, the projected increases in all the hydro-climatic variables considered are greater for the mid of the century than for the end of the century. The magnitude of the projected changes varies across the subbasins with the Jimai subcatchment showing larger changes than Maqu and Tangnag subcatchments. It is also noticed that the magnitude of the projected changes are different under different emission scenarios and GCMs, indicating the uncertainty involved in the impact analysis.

7.1. Introduction

Mountain regions play a vital role for local and downstream water related activities (e.g. Viviroli et al., 2011). These regions are likely to be more vulnerable in the future because of their relatively high sensitivity to climate change. This is particularly true for the Yellow River source region (YRSR), which contributes about 35 % of the total annual runoff of the entire Yellow River and is called the 'water tower' of the Yellow River; a river of critical importance in terms of water resources for China. Historical observations indicate recent warming and drying over the YRSR in recent decades (Zhao et al. 2007; Xu et al. 2007; Hu et al. 2012). This has resulted in a general tendency of decreasing runoff across the region (Zheng et al. 2007; Hu et al. 2011; Cuo et al. 2013). Reduction of rainfall and runoff in recent years across the YRSR has drawn attention about climate change impacts on water resources and its availability in the region. There exists a fairly large body of literature on the climatic impact on the YRSR, but most of the existing studies focused on historical trends of streamflow. To date, very few studies have addressed the future water availability in this region under a changing climate. Xu et al. (2009) applied a hydrological model (SWAT) to investigate the response of streamflow to climate change in the headwater catchment of the Yellow River basin under the emission scenario B2 based on four GCMs outputs. They reported an overall decreasing trend in mean annual streamflow throughout the 21st century. The recent study by Immerzeel et al. (2010) using five GCMs, however, projected a notable 9.5% increase in upstream water yield in the Yellow River under the scenario A1B for the period 2046-2065. Both of the studies have focused on changes in mean streamflow, a detailed exploration of the changes in hydrological extremes (e.g. floods and droughts) and in other hydrological parameters (e.g. evaporation and soil moisture) is lacking to date. This lack of studies contrasts with the relevance of the extreme events for the society and the key role of evaporation and soil moisture in hydrological cycle. Additionally, hydrologic changes

---

[6] This chapter is based on paper Impacts of climate change on the hydrology of the Yellow River source region, China by Hu, Y., Maskey, S. and Uhlenbrook, S. 2014. Submitted to Climatic Change.

in a basin depend on basin characteristics and climatic conditions (Shrestha et al. 2012). The changes could be highly variable for a large mountainous river basin like the YRSR due to the highly heterogeneous physiographic and climatological settings. Understanding future hydrologic changes in a spatially-explicit way is important for water resources management in a large river basin.

In summary, there is (i) an urgent need of understanding spatial and temporal variability of future climate change impacts in the YRSR, (ii) a research gap in assessing climate change impacts on hydrological extremes and other important hydrological parameters in the YRSR and (iii) large uncertainty involved in the climate projections. In this study we aim to tackle these challenges. Thus, the main objectives of this study are to (i) assess possible impacts of climate change on the hydrology (i.e. discharge, evaporation, soil moisture, and snow water equivalent) with particular emphasis on high and low flows in the subbasins of the YRSR and (ii) to assess the range of uncertainty in the simulation of climate change impacts using an ensemble forecast composed of two GCMs and three emission scenarios.

For a spatially distributed evaluation of hydrologic responses, the fully distributed, physically based Water Balance Simulation Model (WaSiM, Schulla 2012) was employed. The WaSiM model was used to simulate the current (1961-1990) and future (2046-2065; 2081-2100) hydrologic regimes based on climate forcings derived from the driving GCMs under the IPCC SRES A2, A1B and B1 emission scenarios. Here, six climate projections consisting of two GCMs and three emission scenarios were used to identify a range of possible hydrologic changes and quantify uncertainties stemming from different GCMs and emission scenarios. For a spatially distributed representation of the climate variables, the Statistical Downscaling Model (SDSM) (Wilby et al. 2002) and the non-homogeneous Hidden Markov Model (NHMM) (Hughes and Guttorp 1994) were used to downscale the GCM outputs to daily mean temperature and daily precipitation, respectively. Both the SDSM and the NHMM have been applied to produce high-resolution climate change scenarios in a range of geographical contexts (Wetterhall et al. 2006; Hu et al. 2013a, b; Xu et al. 2009; Hughes et al. 1999).

7.2. Material and methods

7.2.1. Data set

Daily streamflow data from 3 gauging stations (namely Jimai, Maqu and Tangnag located along the main stream from upstream to downstream) for the period 1961-1990 were collected from YRCC. Daily precipitation at 14 stations, daily mean temperature at 13 stations, and daily wind speed, sunshine, humidity at 7 stations for the same period were obtained from YRCC and CMA. Figure 2.1 depicts the geographical location of the stations in the study region and Table 2.2 shows their latitude, longitude and altitude. Before the analyses, the daily time series were checked for completeness and validated to identify and rectify sequences of anomalous data. The homogeneity of the climatic data was tested by applying the double mass curve method on a monthly basis for each station (Hu et al. 2012). There is missing observation data in the streamflow data for the Jimai station in the year 1990, which was excluded from the study. Slightly less than 0.03% and 0.0017% of the data from two stations for rainfall and temperature, respectively, were missing, which were infilled using the records from neighbouring stations by applying regression functions..

In addition to the observed data, large-scale atmospheric predictors derived from the National Center for Environmental Prediction/National Centre for Atmospheric Research (NCEP/NCAR) reanalysis data set (Kalnay et al., 1996) on a $2.5° \times 2.5°$ grid over the same time period as the observation data were employed for calibration and validation of the statistical downscaling models. These variables include specific humidity, air temperature, zonal and meridional wind speeds at various pressure levels and mean sea level pressure.

In order to project future scenarios, we used outputs from two GCMs (CGCM3.1 (T47) and ECHAM5) for three Emissions Scenarios (A2, A1B and B1) defined by the Intergovernmental Panel on Climate Change (IPCC, 2000). The GCM simulations corresponding to the present (1961–1990) and two future climates (2046–2065 and 2081–2100) were considered in the analysis. Prior to downscalling, both GCM outputs were resampled to 2.5°×2.5° grids fitting the NCEP reanalysis data.

7.2.2. Statistical downscaling models and set up

In this study, the downscaling focues on daily mean temperature and daily rainfall; all other meteorological data (wind speed, relative humidity, relative sunshine duration) remained unchanged. The downscaling is based on the SDSM for temperature and NHMM for precipitation. Both downscaling models belong to the family of stochastic downscaling models. They mainly differ in the way their weather generator parameters are conditioned on large-scale predictors or weather states (Hu et al. 2013a). In SDSM, the multiple linear regression method is used to condition its weather generator parameters on large-scale predictors, whereas in NHMM, this is done using a weather state approach. In addition, SDSM is a single-site model while NHMM are multi-site models. The SDSM can be used to downscale different climatic variables such as precipitation, temperature, humidity and solar radiation, etc., while the NHMM can only be used to downscale precipitation. The NHMM is chosen here to downscale precipitation because of its relatively good performance in reproducing the spatial correlation structure, inter-annual variability and magnitude of the observed precipitation (Hu et al. 2013a). For futher details on the SDSM and the NHMM see Wilby et al. (2003), Wilby and Dawson (2013), Hughes and Guttorp (1994), and Hughes et al. (1999).

Daily atmospheric predictors from the NCEP/NCAR reanalysis and the in-situ daily mean temperature (13 stations) and daily rainfall (14 stations) over the 1961–1980 period are used to develop downscaling relationships. The calibration of the both downscaling models and the predictors selected are described in more detail in Hu et al. (2013a, b). The downscaled-GCM forcings consisting of daily precipitation and temperature, along with other observed climatic varibles were used as inputs to the hydrologic model for the transiet run from 1961-2100. As both SDSM and NHMM are stochastic models, we derived a number of realizations (ensemble) of daily temperature and precipitation. Then to run with the computationally intensive, fully distributed hydrological model, we chose an ensemble member with the least deviation from the ensemble mean.

7.2.3. Hydrologic model and set up

Model description

The Water Balance Simulation Model (WaSiM; Schulla 2012) was employed to simulate the hydrologic response. WaSiM is a physically based, distributed hydrologic model that was originally developed for the quantification of climate change effects in mountainous catchments (Schulla 1997). WaSiM was chosen because of its successful application for modeling hydrologic responses to climate change in several high mountain catchments (Jasper et al. 2006; Bürger et al. 2011; Rössler et al. 2012). The model uses spatial data of topography, land cover, and soil properties combined with interpolations of meteorological point data to calculate the hydrological flux and storage at each raster cell (Rössler et al. 2012). WaSiM is available in different versions. To represent the vertical processes in the unsaturated zone in a physically based way, the Richards equation-based version 9.2.0 was used in this study. The potential evaporation was computed using the Penman–Monteith method (Monteith 1975). Actual evaporation reduces the potential evaporation by a function dependant on the soil moisture status (for more details, see Jasper et al. 2006). The calculation of interception is based on a storage based approach, while storage capacity is linked to the area of the plant surface (leaf area index). Snow accumulation and melt was modeled using the degree-day factor; the same concept is applied to the ice, firn, and snow melt on glacier, but corrected for radiation intensity. Infiltration of water into the unsaturated zone is calculated based on the approach by Green and Ampt (1911), while vertical soil water flow within a defined number of soil layers and depth in the unsaturated zone are described by solving the Richards equation (Richards 1931). In WaSiM, each soil profile is split into several (numerical) layers that may consist of different soil properties (soil layers). At the border of these different soil layers, interflow may be generated. Dependence of the suction head and the hydraulic conductivity on soil moisture content was parameterized according to van Genuchten (1980). Surface flow is the sum of infiltration excess, saturation excess and a defined fraction of the snow melt. Interflow was calculated in different soil layers, depending on suction, drainable water content and saturated hydraulic conductivity. For groundwater modeling we used a conceptual single-linear-storage approach, because more detailed description of the groundwater component was not feasible due to the lack of data in this mountainous catchment. The conceptual groundwater model assumes a permanent soil water exchange (defined by moisture content and permeability) between unsaturated soils and the groundwater table. For each raster cell base flow is derived from the level of the groundwater table using the following equation:

$$Q_b = Q_0 K_s \exp\left[\frac{(h_{GW}-h_{alt})}{k_b}\right] \tag{7.1}$$

where Q_b is the base flow (m/s), Q_0 is the scaling factor for base flow (or maximum base flow if the soil is saturated) (-), K_s is the saturated hydraulic conductivity (m/s), h_{GW} is the height of groundwater table (m a.s.l.), h_{alt} is the altitude of the raster cell (m a.s.l.), and k_b is the recession constant for base flow (m). This approach is applied to each cell of model grid, base flow is thus not generated only at river cells. The parameters Q_0 and k_b have to be calibrated.

In WaSiM, the kinematic wave approach is used in combination with a single linear storage for discharge routing. The approach neglects effects of inertia and diffusion. It is a kinematic wave approach using different flow velocities for different water levels in the channel. After the translation of the wave a single linear storage is applied to the routed discharge in order to consider the effects of diffusion and retardation (Schulla 2012). Then, the discharges from different subbasins are superposed. For a detailed model description, we refer to Schulla (2012).

Model set-up

The model is set up at a spatial resolution of 1 km and run on a daily time step. The following spatial input data were used: (1) Digital Elevation Model based on the Shuttle Radar Topographic Mission (SRTM), version 4 (~90 m resolution; http://srtm.csi.cgiar.org/), (2) land use data based on the Collection 5 MODIS Global Land Cover Type product (~500 m resolution; http://lpdaac.usgs.gov/), and (3) soil data based on the Harmonized World Soil Data base (version 1.2) (~1 km resolution; http://webarchive.iiasa.ac.at/Research/LUC/External-world-soil-database/HTML). All spatial data sets were re-sampled to a resolution of 1 km. Soil hydraulic parameters were obtained from Carsel and Parrish (1998).

Land cover parameters were derived from Schulla (2012) (e.g. root depth, vegetation coverage). In addition to the spatial data described above, WaSiM requires spatially distributed input data time series of temperature, precipitation, relative humidity, relative sunshine duration, and wind speed, supporting the calculation of potential evaporation via the Penman–Monteith method.

Temperature and precipitation were spatially interpolated using a simple nearest-neighbor technique combined with a lapse rate approach. Interpolation of other meteorological input data was done by inverse distance weighting. WaSiM set up does not provide the option of specifying temporally variable lapse rates in its current set-up, constant lapse rates throughtout the year were applied in this study. A fixed temperature lapse rate of -6.5°C/km was derived based on regressing the mean annual temperature of the studied sations against their elevation. This temperature lapse rate was comparable with those used in the previous studies in the neighboring mountain region (Zhang et al. 2008b; Liu and Chen 2000; Schaner et al. 2012; Immerzeel et al. 2012a; Immerzeel et al. 2012b). Based on our understanding of the distribution of precipitation over the catchment, we used two separate precipitation lapse rates for elevations below and above 5000 m asl. Earlier studies reported a peak in precipitation between 5000 and 6000 m asl in the Himalayas (Immerzeel et al. 2012b; Young and Hewitt 1990). Immerzeel et al. (2012b) assumed that precipitation increases linearly to an elevation of 5500 m asl and decreases with the same rate at higher elevation in the upstream part of the Indus. Based on regression analysis, the annual precipitation lapse rate of 87 mm/km was derived in this study. However, as precipitation lapse rates are applied over the daily precipitation data in WaSiM, the daily precipitation lapse rate is estimated as the annual precipitation lapse rate divided by the number of rainy day in a year. Because most stations are located below 5000 m asl, we assume that precipitation decreases with the same rate at elevations above 5000 m asl. Hence, the daily lapse rates of 0.7 mm/km and -0.7 mm/km were used for precipitation at elevations below and above 5000 m asl (note that areas with elevation above 5000 m asl are less than 1% of the catchment), respectively. A similar precipitation lapse rate was also reported by Zhang et al. (2008b) for the same region.

The model was calibrated against observed discharges at Tangnag gauging station for the time period 1965-1984 and validated for the period 1985-1990. The period 1961-1964 was used to initialize the model. We carried out manual calibration of the model on following most sensitive parameters: 1) recession constant of direct runoff (k_d), 2) recession constant of interflow (k_i), 3) interflow drainage density (d_r), 4) recession constant of base flow (k_b), 5) scaling factor for base flow (Q_0), and 6) recession constant for the saturated hydraulic conductivity ($krec$). Schulla (2012) found discharge simulation to be most sensitive to $krec$ and d_r. To examine the WaSiM model performance for the calibration and validation periods, we used two statistical criteria: the Nash-Sutcliffe effiency (NSE, Nash and Sutcliffe 1970) and the coefficient of determination (R^2).

7.2.4. Climate change impact detection indices

We investigated changes in hydrologic regimes from the control period (1961-1990) for the two future periods (2046–2065 and 2081–2100) using four indices. These indices represent relevant responses of the hydrological system to changes in temperature and precipitation:

(1) Changes in annual and seasonal discharge, potential and actual evaporation, and soil moisture;
(2) Snow storage (April 1st snow water equivalent (SWE));
(3) Annual peak discharge and 7-day low flows; and
(4) Timing of annual peak discharge.

7.3. Results and discussions

7.3.1. WaSiM calibration and validation

Overall, the WaSiM model is able to reproduce the runoff dynamics for both the calibration and validation periods with NSE and R^2 greater than 0.8 for the three subbasins. Figure 7.1 shows a comparison between observed and simulated daily flow at Jimai, Maqu and Tangnag gauging stations during the validation period. As can be seen from Figure 7.1, there is a good agreement between the observed and simulated discharge at three gauging stations as also indicated by higher values of NSE>0.8 and R^2>0.83. The model performance is generally better for the downstream Maqu and Tangnag cathments than the upstream Jimai catchment. This may be partly due to problems in the meteorological input data, especially at higher elevations where only few climate stations are located and accurrate measurements are increasingly difficult. The resulting set of optimal parameters is presented in Table 7.1.

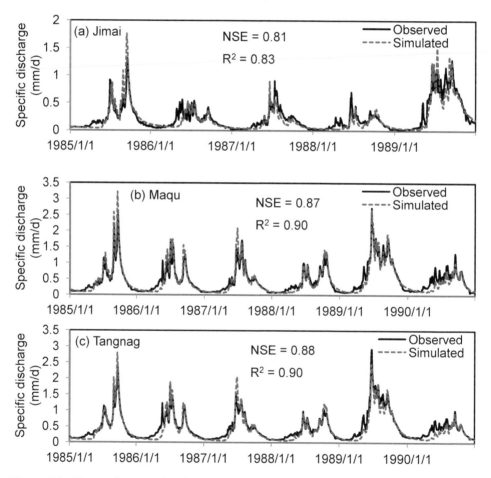

Figure 7.1: Observed versus simulated daily discharge at (a) Jimai station, (b) Maqu station and (c) Tangnag station during the validation period (1985-1990). Jimai is only validated for 1985-1989 because of missing observation data in 1990.

Table 7.1: Parameters of the WaSiM model and their optimal values resulting from calibration (1965-1984) and validation (1985-1990)

Parameter	Symbol	Unit	Optimal value
Recession constant for direct runoff	k_d	[h]	30
Recession constant for interflow	k_i	[h]	60
Drainage density	d_r	[m^{-1}]	20
Base flow recession constant	k_b	[m]	0.6
Scaling factor for base flow	Q_0	[-]	0.6
Recession constant for the saturated hydraulic conductivity	k_{rec}	[-]	0.1

7.3.2. Changes in temperature and precipitation

The results presented in the following sections are analyzed on annual basis and for two seasons: wet season (June-September) and dry season (October and May). Projected changes for the two future periods (2046–2065 and 2081–2100) are investigated against the control period (1961–1990). The changes in the mean correspond to the difference or percentage difference between the mean values in the future period and those in the control period. A two-tailed Student's t test for the 5% confidence level was performed to check if the mean values from the present and future periods are significantly different.

Figure 7.2 shows changes in temperature and precipitation for the two GCMs and the three emission scenarios. All of the projections show statistically significant increases in both temperature and precipitation throughout the year. As expected, the projected increases for 2081-2100 are generally greater than those for 2046-2065. Additionally, the range of seasonal and annual variations of precipitation and temperature are higher for 2081-2100 than for 2046-2065. The seasonal variations of precipitation and temperature are higher in the dry season than in the wet season. Also, we note that the projected temperature increases in 2081–2100 tend to scale with the emission scenario, i.e. the larger the greenhouse gas forcing, the stronger the response (generally most intense in the A2, followed by the A1B and B1 scenarios). However, the same does not hold true for 2046–2065, where in some cases the projected changes for the middle of the 21st century are stronger in the A1B scenario than in the A2 scenario. Similar findings were also reported by Hu et al. (2013b) for other temperature indices based on same GCMs, emission scenarios and statistical downscaling model. The ECHAM5-driven projections show larger temperature increases than the CGCM3-driven ones, in particular in 2081-2100, while the reverse is found for precipitation increases in 2046-2065. For precipitation in 2081-2100, the ECHAM5 shows larger wet season and annual increases than CGCM3 under the A2 and B1 scenarios while the opposite is observed for the dry season. The annual mean temperature is projected to increase by 2.8-2.9°C for A2, by 2.6-3.3°C for A1B and by 2.1-2.6°C in 2046-2065 and by 4.7-6.0°C for A2, by 3.7-5.4°C for A1B and by 2.5-4.0°C in 2081-2100. Seasonally, the ECHAM5-driven projections show greater temperature increases in dry season than in wet season while the CGCM3-driven ones show comparable temperature increases between dry and wet seasons. Regarding precipitation, annual precipitation is projected to increase by 12.8-15.4% for A2, by 16.1-19.0% for A1B and by 11.3-14.7% for B1 in 2046-2065 and by 26.8-30.3% for A2, by 19.2-25.0% for A1B and by 18.5-21.4% for B1 in 2081-2100. Percentage changes in the dry season precipitation are generally greater than those in wet season.

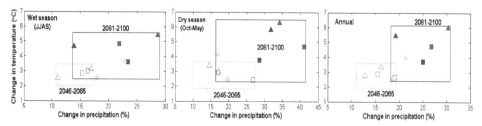

Figure 7.2: Projected average changes in temperature and precipitation according to two global climate models CGCM3 (square) and ECHAM5 (trangle) and three emission scenarios A2 (red), A1B (blue) and B1 (green) for 2046-2065 (open symbols) and 2081-2100 (closed symbols).

7.3.3. Changes in streamflow

Figure 7.3 shows projected changes in annual, seasonal, high and low flows from the two GCMs and three emission scenarios for the two future periods. As shown in Figure 7.3(a), all projections suggest statistically significant increases in annual flow for both future periods. Annual flow is expected to increase by 18.3-20.6% for A2, by 22.6-28.4% for A1B and by 18.2-21.7% for B1 in 2046-2065 and by 34.5-40.4% for A2, by 18.9-36.0% for A1B and by 28.6-33.0% for B1 in 2081-2100. Similar results are obtained for seasonal flow with the CGCM3 (ECHAM5) showing greater increases in wet (dry) season flow. High flow is projected to significantly increase in all the projections with the exception of the ECHAM5-A1B driven projection in 2081-2100, while the smallest increases are projected for low flow with the CGCM3 showing insignificant increases in 2046-2065 under the A2 and B1 scenarios and in 2081-2100 under the B1 scenario. Compared to significant increases in the magnitude of high flow, the timing of high flow shows insignificant increases in most projections (Figure 7.3(f)). In general, a later occurrence of peak discharge is simulated in most projections, which can be explained as the effect of the low dependence on meltwater and the increased precipitation during the wet summer. In general, the runoff increases are larger for the Jimai subcatchment than at the subcatchments Maqu and Tangang. The increases in streamflow are mainly due to increased precipitation. These results are consistent with previous studies, which have shown that changes in streamflow are mainly driven by changes in precipitation (Hu et al. 2011). Projected increases in high flow are probably as a result of increased heavy precipitation events reported in an earlier study by Hu et al. (2013a).

Note that the projected increases in streamflow in this study are at odds with the study by Xu et al. (2009), which suggested an overall decrease in annual streamflow for this region for the 21st century under the B2 scenario using two downscaling methods (the SDSM and the delta-change) and four GCMs (CGCM2, CCSR, CSIRO and HadCM3). This discrepancy is probably due to choice of different GCMs or different downscaling methods. However, our results agree well with studies performed in the the upstream or the entire Yellow River (Immerzeel et al. 2010; Liu et al. 2011). Using the outputs from five different GCMs under the A1B scenario, Immerzeel et al. (2010) reported a notable 9.5% increase in upsteam water yield and earlier onset of snowmelt peak in the Yellow River for the period 2046-2065 relative to 2000-2007. Using two downscaling methods (the SDSM and the combination of bilinear-interpolation and delta-change) and one GCM (HadCM3), Liu et al. (2011) suggested an increase in annual streamflow for the entire Yellow River for the 21st century under the A2 and B2 scenarios.

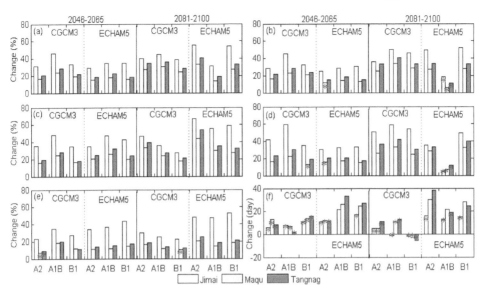

Figure 7.3: Projected changes in (a) annual, (b) wet season, (c) dry season, (d) high, (e) low flows, and (f) the timing of annual peak flow at Jimai, Maqu and Tangnag gauging stations for the two future periods 2046-2065 and 2081-2100 relative to baseline period 1961-1990. Asterisks represent changes that are not statistically significant at 0.05 significant level.

7.3.4. Changes in evaporation

As shown in Figure 7.4, potential and actual evaporation (E_{pot} and E_{act}) are expected to increase significantly through the entire year for both future periods. The changes in E_{pot} and E_{act} are comparable between the three studied catchments. For the Tangnag catchment, increases in annual E_{act} range from 9.5-10.0% for A2, 8.4-11.5% for A1B and 8.4-8.5% for B1 in 2046-2065 and by 18.5-23.2% for A2, by 13.6-19.4% for A1B and by 9.8-14.8% for B1 in 2081-2100. Increases in E_{act} are smaller than the corresponding increases in E_{pot}. Seasonally, the ECHAM5 driven projections show larger evaporation increases in the dry season than in the wet season, which reflects the seasonality of projected temperature increases by this model. As in the case of the temperature, projected increases in both E_{pot} and E_{act} in 2081–2100 are generally most intense in the A2, followed by the A1B and B1 scenarios. This indicates that the increases in evaporation are probably mainly due to increased air temperature.

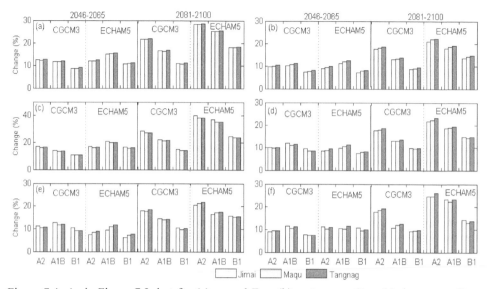

Figure 7.4: As in Figure 7.3, but for (a) annual E_{pot}, (b) wet season E_{pot}, (c) dry season E_{pot}, (d) annual E_{act}, (e) wet season E_{act} and (f) dry season E_{act}.

7.3.5. Changes in soil moisture and snow storage

All the projections show significant increases in soil moisture (SM, root zone soil water content) on annual basis ranging from 2.8 to 5.2% in 2046-2065 and 3.0 to 5.4% in 2081-2100 for the Tangang catchment (Figure 7.5a). Although the increases in soil moisture are also projected on seasonal basis, but only half of the projected changes are statistically significant. The CGCM3 shows insignificant increases in soil moisture during the dry season while similar results are obtained for the ECHAM5 during the wet seasons. As in the case of the runoff changes, the increases in soil moisture are generally greater in the Jimai catchment. The increases in soil moisture are mainly due to increased precipitation that exceeds the increase in AET. The potential impacts of climate change on snow storage were considered for the WaSiM simulated April 1st snow water equivalent (SWE) anomalies. The available SWE on April 1st is an indication of the amount of snowpack storage that will eventually melt and generate runoff, or infiltrate and be stored in shallow groundwater reserves and released over the summer (Bennett et al. 2012). The potential future change in the snow storage is evident in the April 1st SWE anomalies, which predict significant declines for most scenarios (Figure 7.5d). The decline in April 1st SWE of the Tangnag catchment ranges from -21.7 to -72.8% in 2046-2065 and -54.8 to -85.2% in 2081-2100 with the ECHAM5-driven projections showing more pronounced decline than the CGCM3-driven ones. This decline is likely a combined result of increase in precipitation falling as rainfall in winter and earlier snowmelt caused by rising temperature.

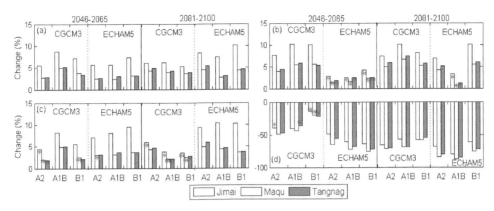

Figure 7.5: As in Figure 7.3, but for (a) annual SM, (b) wet season SM, (c) dry season SM and SWE.

7.4. Conclusions

All climate projections for the mid and the end of the 21st century indicate substantial warmer and wetter climate over the study region. Under the warmer and wetter climate projected for the 21st century, the Yellow River source region is expected to experience significant increases in annual and seasonal streamflow, accompanied by strong increases in high flows and small increases in low flows. The analysis of the changes in snow storage suggests a significant reduction in April 1st SWE in most scenarios. The peak flows are likely to occur later in the future in this rainfall-dominated basin. Due to increased temperature, enhanced evaporation is projected on both annual and seasonal basis. Projected increases in soil moisture are significant on annual basis, but the predicted changes on the seasonal scale are insignificant. Overall, the projected increases in all the hydro-climatic variables considered tend to increase with time. The magnitude of the projected changes varies across the subbasins with Jimai catchment showing larger changes than the Maqu and Tangnag subcatchments.

The anticipated changes in the hydrology of YRSR are likely to have significant implications for water resources management in the basin. On the one hand, the expected increases in annual and seasonal runoff, if properly managed, could yield a positive effect on water avaliablility for this region and downstream water supply and environmental flows, and alleviate current water shortage in the Yellow River basin. On the other hand, the increasing peak flows could pose an increased flood risk. At the same time, other natural hazards such as soil erosion and landslides are expected to exacerbate owing to the increasing heavy rainfall events and flooding. Current water management practices and strategies need to be reassessed and consider projected changes in both mean and extreme flows and uncertainties therein.

As the first comprehensive study on the hydrologic impacts of climate change for the YRSR, our results demonstrate a need for integrating various sources of uncertainty into climate change impact studies. The simulated hydrologic impacts of climate change are subject to large uncertainties related to the emission scenarios, GCMs, downscaling method and hydrologic model, but also the changes in land use/cover. First, the uncertaity associated with the choice of downscaling techniques was neglected here due to the use of only one precipitation (temperature) downscaling model. Second, due to a relatively small number of GCMs and emission scenarios applied, only a limited estimation of the possible uncertainty associated with such scenarios could be quantified. Third, in this study land cover/land use is assumed static throughout the projection timeframe (over the 100 years) given the largerly unporpulated and relatively pristine catchment considered here. Climate change, however, can alter the vegetation pattern and thus have an effect on hydrology. A more comprehensive uncertainty analysis identifying the role of emission scenarios, climate models and downscaling approaches is recommended for further studies of climate change impacts in the region. Furthermore, future studies should consider the combined effects of climate change and land use/cover change, and associated feedbacks between them.

8. Synthesis, conclusions and recommendations

8.1. Historical hydroclimatic variability and their linkages

In the latter half of the twentieth century the YRSR has become warmer and experienced some seasonally varying changes in rainfall. This warming is mainly attributed to the increase in the minimum temperature as a result of the increase in the magnitude and decrease in frequency of low temperature events. In contrast to the temperature indices, the trends in rainfall indices are less distinct. However, on a basin scale increasing trends are observed in winter and spring rainfall. Conversely, the frequency and contribution of moderately heavy rainfall events to total rainfall show a significant decreasing trend in summer. However, it is unclear as to whether these trends are part of a longer period of oscillation or the result of long term climate change.

Trends in hydrological regime vary considerably from one subbasin to another. Overall, the YRSR has been characterized by an overall tendency towards decreasing water availability. The hydrological variables studied are closely related to precipitation in the wet season (June, July, August and September), indicating that the widespread decrease in wet season precipitation is expected to be associated with decrease in streamflow. To conclude, decreasing precipitation, particularly in the wet season, along with increasing temperature can be associated with pronounced decrease in water resources. However, note that the observed precipitation and streamflow changes are inconsistent with future projections presented in the following sections. The likely reasons behind the inconsistency are briefly discussed in section 8.5.

8.2. Future rainfall scenarios derived from different downscaling techniques

Three statistical downscaling methods (SDSM, GLIMCLIM and NHMM) are compared with regard to their ability to downscale summer (June–September) daily precipitation over the YRSR. In comparison with other two models, NHMM shows better performance in reproducing the spatial correlation structure, inter-annual variability and magnitude of the observed precipitation. However, it shows difficulty in reproducing observed wet- and dry spell length distributions at some stations. SDSM and GLIMCLIM showed better performance in reproducing the temporal dependence than NHMM. These models are also applied to derive future scenarios for six precipitation indices for the period 2046–2065 using the predictors from two global climate models (GCMs; CGCM3 and ECHAM5) under the IPCC SRES A2, A1B and B1scenarios. There is a strong consensus among two GCMs, three downscaling methods and three emission scenarios in the precipitation change signal. Under the future climate scenarios considered, all parts of the study region would experience increases in rainfall totals and extremes that are statistically significant at most stations. The magnitude of the projected changes is more intense for the SDSM than for the other two models. The increase in the magnitude of rainfall totals and extremes is also accompanied by an increase in their inter-annual variability. Overall, this study highlights the importance of acknowledging limitations and advantages of different statistical downscaling methods, and also implies that climate projection based on only one GCM or one downscaling model should be interpreted with caution.

8.3. Future temperature changes and elevation dependency

Using the Statistical DownScalingModel (SDSM) and the outputs from two global climate models, we investigate possible changes in mean and extreme temperature indices and their elevation dependency over the YRSR for the two future periods (2046–2065 and 2081–2100) under three IPCC SRES emission scenarios (A2, A1B and B1). For the middle and end of the 21st century all parts of the study region may experience increases in both mean and extreme temperature in all seasons, along with an increase in the frequency of hot days and warm nights and decrease in frost days. As expected, the projected changes in the temperature indices in 2081–2100 are generally larger than those in 2046–2065. By the end of the 21st century (2081-2100), inter-annual variability increases in the frequency of hot days and warm nights in all seasons. The frost days show decreasing inter-annual variability in spring and increasing one in summer. Several indices demonstrate elevation-dependent changes, which varies from index to index and from season to season. For autumn pronounced elevation-dependent changes are observed in which around six out of eight indices show significant increasing changes with elevation.

8.4. Spatial and temporal variability of future hydrologic impacts of climate change

Using the statistically downscaled outputs from two GCMs and a fully distributed, physically based hydrologic model (WaSiM), we investigated potential changes in the future hydrologic regimes of the YRSR. The results revealed that a warmer and wetter climate is likely to bring considerable hydrologic changes in the YRSR, where it is expected to experience significant increases in annual and seasonal streamflow, accompanied by strong increases in high flow and small increases in low flow. This may yield a positive effect on water avaliablility for this region and downstream water supply. The analysis of changes in snow storage suggests a significant reduction in April 1st SWE in most scenarios. The peak flow is likely to occur later in the future in this rainfall-dominated basin. Due to increased temperature, enhanced potential and actual evaporation is projected throughout the year. Projected increases in soil moisture are significant on annual basis but seasonally insignificant. Overall, the projected increases in all the hydro-climatic variables considered are greater for the mid of the century than for the end of the century. The magnitude of the projected changes varies across the subbasins with Jimai catchment showing larger changes than Maqu and Tangnag catchments. It is also noticed that the magnitude of the projected changes are different under different emission scenarios and GCMs, indicating the uncertainty involved in the impact analysis. A strong limitation to all the results presented in the present study is the assumption of a static vegetation cover. However, vegetation is very likely to change with a changing climate, especially in a warmer and wetter climate, and thus have an effect on hydrology. Such changes were not included in our study. Therefore, future studies should consider the combined effects of climate change and land use/cover change, and associated feedbacks between them.

The anticipated changes in the hydrology of YRSR could have significant implications for water resources management in the basin. On the one hand, the expected increases in annual and seasonal runoff, if properly managed, could yield a positive effect on water avaliablility for this region and downstream water supply, and alleviate current water shortage in the Yellow River basin. On the other hand, the increasing peak flows could pose an increased flood risk. At the same time, other natural hazards such as soil erosion and landslide are expected to exacerbate owing to the increasing extreme rainfall events and flooding. Furthermore, the differences in the future hydrologic responses across the basin also support the need for a spatial and temporal evaluation of future hydrologic change. Such an approach is especially relevant for large river basin like the Yellow River, where physiographic and climatic characteristics vary considerably. Moreover, the spatial and temporal evaluation of future responses will allow for consideration of adaptation strategies more suited for local conditions. Current water management practices and strategies (e.g.

reservoir operation) may need to be redesigned to consider projected changes in both mean and extreme flows and uncertainties therein.

8.5. Consistency of observed and projected hydroclimatic changes

By linking past changes to expected future changes in the YRSR, we find that the recently observed warming is consistent with regional climate change projections, indicating that the recently observed warming is a plausible illustration of future expected warming in the region. By contrast, we find a notable discrepancy in precipitation where observations show less distinct trends while future projections show significant year-round increases. The most striking inconsistency is the contradiction between the projected increase and the observed decrease in streamflow. Reasons to explain the observed inconsistency are manifold and we will highlight the only two important ones. A detailed analysis of the real reasons for the inconsistency, however, is beyond the scope of this thesis. First, the trends in precipitation and streamflow observation data may be mainly due to natural (internal) variability. The high natural variability of the YRSR climate in both space and time leads to low signal-to-noise ratio of externally forced (anthropogenic) changes. The observed trends contain large signals related to the North Atlantic Oscillation (NAO) and El Niño-Southern Oscillation (ENSO), of which a major unknown part may be unrelated to the anthropogenic signal (Cuo et al., 2013). However, the anthropogenic factors such as the greenhouse gas emission and sulfate concentrations are the dominant forcing for future precipitation and streamflow changes. Second, the inconsistency could also be due to the misrepresentation of some processes relevant for precipitation formulation in the current climate model such as topographic forcing and the influence of the NAO and ENSO.

Overall, communication of future expected change of precipitation and streamflow is complicated by the fact that the expected future changes are inconsistent with observed changes. The detection of an outright sign-reversal in the observed and projected trends provides strong evidence that the recent observed changes cannot be used to illustrate the future expected changes of streamflow. Such inconsistency calls for an urgent need for research aiming to reconcile the historical changes with future projections.

8.6. Limitations and recommendations

The presented scenarios of hydroclimatic changes and the associated methodology give rise to further discussion and research. It is important to keep in mind a few caveats which were not addressed in this study. First, in this study the land cover is assumed static throughout the projection timeframe. Climate change, however, can alter the vegetation pattern and thus have an effect on hydrology. Such changes were not included in our study. Therfore, one important future research area is to incoporate explicitly land cover changes which are taking place at the same time as climate change and the feedbacks between climate change and land cover change. In many places, especially in mountainous areas, these combined effects are more important than only looking at changes in rainfall and temperature (Viviroli et al., 2011).

Second, the present study examined uncertainties arising from a relatively small sample of emission scenarios and climate models. Uncertainties due to downscaling methods and hydrological model parameterisations and structures were not considered. The results of this study in conjunction with many published studies indicate that uncertainties occur in each step involved in a climate change impact assessment. Hence, a more comprehensive analysis of impact uncertainty should be performed by applying multimodel ensemble simulations consisting of several emission scenarios, climate models and downscaling approaches.

Third, assessment of water resources from remote mountainous catchments plays a crucial role for the development of downstream areas in or in the vicinity of mountain ranges. The YRSR is a crucial area in terms of water resources, but our understanding of the response of its high-elevation catchments to a changing climate is hindered by lack of hydro-meteorological and cryospheric data. Climate and hydrological modeling is particularly challenging here because of the complex physiography and hydroclimatic system. Data scarcity adds to this difficulty by preventing the application of systematic calibration procedures that would allow better identification of the parameter sets. Remote sensing provides objective and quasi-continuous information on relevant variables and could provide a solution for this issue. The use of remote sensing in climate and hydrological modelling is a growing field and proves to be highly relevant, especially in areas where data are scarce and unreliable (Maskey et al., 2011; Bastiaanssen et al., 2011, Dente et al., 2012). For future applications the combined use of remotely sensed and insitu data has the potential to hydrological modelling research in this remote mountainous catchment. However, the importance of maintaining and strengthening existing hydro-meteorological networks, by increasing the number of measuring locations and extending the length of records, should not be neglected, even if remotely sensed data sets are becoming increasingly avaliable to hydrological modeling studies. Remotely sensed data might serve to reduce error and uncertainty in hydrological models for data scarce or ungauged basins but is no substitute for a dense in situ gauge network (Neal et al., 2009). Therefore, another crucial area of future work is to maintain and increase existing hydro-meteorological networks in the YRSR. The limitations of the observational network are clearly evidenced in Figure 2.1. New initiatives are being proposed by YRCC to hydro-climatic monitoring network across the Yellow River. This will help in understanding the reasons for observed changes in climate extremes and in improving confidence and accuracy in projected changes.

References

Abdul Aziz OI, Burn DH (2006) Trends and variability in the hydrological regime of the Mackenize River Basin. Journal of Hydrology 319: 282–294.

Anandhi A, Srinivas VV, Nanjundiah RS, Nagesh Kumar D (2008) Downscaling precipitation to river basin in India for IPCC SRES scenarios using support vector machine. Int J Climatol 28:401-420.

Anandhi A, Srinivas VV, Nagesh Kumar D, Nanjundiah RS (2009) Role of predictors in downscaling surface temperature to river basin in India for IPCC SRES scenarios using support vector machine. Int J Climatol 29:583–603.

Arnell NW (1999) Climate change and global water resources. Global Environmental Change 9: 31–49.

Bastiaanssen WGM, Cheema MJM, Immerzeel WW, Miltenburg I, Pelgrum H (2012) The surface energy balance and actual evapotranspiration of the transboundary Indus Basin estimated from satellite measurements and the ETLook model, Water Resources Research, vol. 48, doi 10.1029/2011 WR 0101482.

Bae DH, Jung IW, Chang H (2008) Long-term trend of precipitation and runoff in Korean river basins. Hydrological Processes 22: 2644–2656.

Bates BC, Kundzewicz ZW, Wu S, Palutikof J (2008) Climate change and water. Technical Paper of the Intergovernmental Panel on Climate Change, IPCC Secretariat, Geneva.

Battisti DS, Naylor RL (2009) Historical warnings of future food insecurity with unprecedented seasonal heat. Science 323(5911): 240–244. DOI: 10.1126/science.1164363.

Biasutti M (2013) Climate change: Future rise in rain inequality. Nature Geoscience 6: 337-338.

Bellone E, Hughes JP, Guttorp P (2000) A hidden Markov model for downscaling synoptic atmospheric patterns to precipitation amounts. Clim Res 15:1–12.

Beniston M (2003) Climatic change in mountain regions: a review of possible impacts. Clim Change 59:5–31.

Beniston M, Diaz HF, Bradley RS (1997) Climatic change at high elevation sites: an overview. Clim Change 36: 233–251.

Beniston M. and Rebetez M (1996) Regional behavior of minimum temperatures in Switzerland for the period 1979–1993, Theor. Appl. Climatol., 53, 231–243.

Birsan MV, Molnar P, Burlando P, Pfaundler M (2005) Streamflow trends in Switzerland. Journal of Hydrology 314: 312–329.

Burn DH (2008) Climatic influences on streamflow timing in the headwaters of the Mackenzie River Basin. Journal of Hydrology 352: 225–238.

Burn DH, Hag Elnur MA (2002) Detection of hydrologic trends and variability. Journal of Hydrology 255: 107–122.

Bürger G, Schulla J,Werner AT (2011) Estimates of future flow, including extremes, of the Columbia River headwaters, Water Resour. Res., 47, W10520, doi:10.1029/2010WR009716.

Brunetti M, Buffoni L, Maugeri M, Nanni T (2000) Precipitation intensity trends in northern Italy. Int J Climatol 20:1017–1031.

BrunettiM, Maugeri M, Nanni T (2001) Changes in total precipitation, rainy days and extreme events in northeastern Italy. Int J Climatol 21:861–871.

Calanca PL, Roesch A, Jasper K, Wild M (2006) Global warming and the summertime evapotranspiration regime of the Alpine region, Clim. Change, 79, 75–78.

Carsel RF, Parrish RS (1988) Developing joint probability distributions of soil water retention characteristics. Water Resour. Res., 24(5), 755–769.

Chandler RE (2002) GLIMCLIM: generalised linear modelling for daily climate series (Software and User Guide). Department of Statistical Science, University College London.

Chandler RE, Wheater HS (2002) Analysis of rainfall variability using generalised linear models—a case study from the west of Ireland. Water Resour Res 38:W1192.

Chen YN, Takeuchi K, Xu CC, Chen YP, Xu ZX (2006) Regional climate change and its effects on river runoff in the Tarim Basin, China. Hydrological Processes 20: 2207–2216.

Chen B, Chao W, Liu X (2003) Enhanced climatic warming in the Tibetan Plateau due to doubling CO2: a model study, Clim.Dynam., 20, 401–413.

Chen D, Chen Y (2003) Association between winter temperature in China and upper air circulation over East Asia revealed by canonical correlation analysis, Global Planet. Change, 37, 315–325.

Chou, C. et al. (2013) Increase in the range between wet and dry season precipitation, Nature Geoscience, 6, 263–267.

Chu J, Xia J, Xu C, Singh V (2010) Statistical downscaling of daily mean temperature, pan evaporation and precipitation for climate change scenarios in Haihe River, China, Theor. Appl. Climatol., 99, 149–161, doi:10.1007/s00704-009-0129-6.

Clark RT, Brown SJ, Murphy JM (2006) Modeling northern hemisphere summer heat extreme changes and their uncertainties using a physics ensemble of climate sensitivity experiments, J. Clim. 19 4418–35.

Cong Z, Yang D, Gao B, Yang H, Hu H (2009) Hydrological trend analysis in the Yellow River basin using a distributed hydrological model. Water Resource Research 45, W00A13, doi: 10.1029/2008WR006852.

Dente L, Vekerdy Z, Wen J, Su Z (2012). Maqu network for validation of satellite-derived soil moisture products. International Journal of Applied Earth Observation and Geoinformation 17, 55–65.

Diaz-Nieto J, Wilby RL (2005) A comparison of statistical downscaling and climate change factor method: impacts on low flows in the River Thames, United Kingdom. Climatic Change 69:245–268.

Diaz HF, Bradley RS (1997) Temperature variations during the last century at high elevation sites, Climatic Change, 36, 253-279.

Diaz HF, Eischeid JK (2007) Disappearing álpine tundrá Köppen climatic type in the western United States, Geophys. Res. Lett., 34, L18707, doi:10.1029/2007GL031253.

Di Baldassarre, G., Elshamy, M., van Griensven, A., Soliman, E., Kigobe, M., Ndomba, P., Mutemi, J., Mutua, F., Moges, S., Xuan, J.-Q., Solomatine, D. & Uhlenbrook, S. Future hydrology and climate in the River Nile basin: a review. Hydrol. Sci. J. 56(2), 2011, 199-211.

Dong XH, Yao ZJ, Chen CY (2007) Runoff variation and its response to precipitation in the source region of the Yellow River. Resources Science 29(3): 67–73 (in Chinese).

Easterling DR, Horton B, Jones PD, Peterson TC, Karl TR, Parker DE, Salinger MJ, Razuvayev V, Plummer N, Jamason P, Folland CK (1997) Maximum and minimum temperature trends for the globe. Sci 277:364–367.

Fan F, Mann ME, Lee S, Evans JL (2010) Observed and modeled changes in the South Asian Summer Monsoon over the historical period, J. Climate, 23, 5193–5205.

Fanta B, Zaake BT, Kachroo RK (2001) A study of variability of annual river flow of the southern African region. Hydrological Sciences 46(4):513–524.

Fealy R, Sweeney J (2007) Statistical downscaling of precipitation for a selection of sites in Ireland employing a generalized linear modelling approach. Int J Climatol 27:2083–2094.

Fischer E, Schär C (2009) Future changes in daily summer temperature variability: driving processes and role for temperature extremes, Clim. Dynam., 33, 917–935.

Fischer EM, Schär C (2010) Consistent geographical patterns of changes in high-impact European heatwaves Nature Geosci.,3, 398–403.

Frost AJ, Charles SP, Timbal B, Chiew FHS, Mehrotra R, Nguyen KC, Chandler RE, McGregor J, Fu G, Kirono DGC, Fernandez E, Kent D (2011) A comparison of multi-site daily rainfall downscaling techniques under Australian conditions. J Hydrol 408:1–18.

Forney GD Jr (1978) The Viterbi algorithm. Proc IEEE 61:268–278

Fu GB, Chen SL, Liu CM, Shepard D (2004) Hydro-climatic trends of the Yellow River basin for the last 50 years. Clim Change 65:149–178.

Fu GB, Barber ME, Chen SL (2009) Hydro-climatic variability and trends in Washington State for the last 50 years. Hydrological Processes 24: 866–878.

Fu GB, Charles SP, Viney RN, Chen SL, Wu JQ (2007) Impacts of climate variability on stream-flow in the Yellow River. Hydrological Processes 21: 3431–3439.

Gao X, Shi Y, Song R et al (2008) Reduction of future monsoon precipitation over China: comparison between a high resolution RCM simulation and the driving GCM. Meteorol Atmos Phys 100:73–86. doi:10.1007/s00703-008-0296-5.

Ghosh S (2010) SVM-PGSL coupled approach for statistical downscaling to predict rainfall from GCM output. J Geophys Res 115: D22102. doi:10.1029/2009JD013548.

Gibson CA, Meyer JL, Poff NL, Hay E, Georgakakos A (2005) Flow regime alterations under changing climate in two river basins: implications for freshwater ecosystems. River Research and Applications 21: 849–864.

Groisman PY, Knight RW, Easterling DR, Karl TR, Hegerl GC, Razuvaev VAN (2005) Trends in intense precipitation in the climate record. J Climate 18:1326–1350.

Green WH, Ampt GA (1911) Studies on soil physics: I. The flow of air and water through soils, J. Agric. Sci., 4, 1–24.

Giorgi F (2006) Climate change hot-spots, Geophys. Res. Lett., 33, L08707, doi:10.1029/2006GL025734.

Giorgi F, Hurrell J, Marinucci M, Beniston M (1997) Elevation dependency of the surface climate change signal: a model study, J. Climate, 10, 288–296.

Ghosh S (2010) SVM-PGSL coupled approach for statistical downscaling to predict rainfall from GCM output, J. Geophys. Res., 115, D22102, doi:10.1029/2009JD013548.

Hannaford J, Marsh TJ (2006) An assessment of trends in UK runoff and low flows using a network of undisturbed catchments. International Journal of Climatology 26: 1237–1253.

Hannaford J, Marsh TJ (2008) High-flow and flood trends in a network of undisturbed catchments in the UK. International Journal Climatology 28: 1325–1338.

Haylock M, Goodess C (2004) Interannual variability of European extreme winter rainfall and links with mean large-scale circulation. Int J Climatol 24:759–776.

Haylock M, Nicholls N (2000) Trends in extreme rainfall indices for an updated high quality data set for Australia, 1910–1998. Int J Climatol 20:1533–1541.

Haylock MR, Cawley GC, Harpham C, Wilby RL, Goodess CM (2006) Downscaling heavy precipitation over the United Kingdom: a comparison of dynamical and statistical methods and their future scenarios, Int. J. Climatol., 26, 1397–1415, doi:10.1002/joc.1318.

Hirsch RW, Slack JR, Smith RA (1982) Techniques of trend analysis for monthly water quality data. Water Resources Management 22: 1159–1171.

Hughes JP, Guttorp P (1994) A class of stochastic models for relating synoptic atmospheric patterns to regional hydrological phenomena, Water Resour. Res., 30(5), 1535–1546.

Hughes JP, Guttorp P, Charles SP (1999) A non-homogeneous hidden Markov model for precipitation occurrence, Appl. Stat., 48(1), 15–30.

Hughes JP, Guttorp P (1994) A class of stochastic models for relating synoptic atmospheric patterns to regional hydrological phenomena. Water Resour Res 30(5):1535–1546.

Hundecha Y, Bardossy A (2005) Trends in daily precipitation and temperature extremes across western Germany in the second half of the 20th century. Int J Climatol 25:1189–1202.

Hundecha Y, Bärdossy A (2008) Statistical downscaling of extremes of daily precipitation and temperature and construction of their future scenarios, Int. J. Climatol., 28, 589–610, doi:10.1002/joc.1563.

Huntington TG (2006) Evidence for intensification of the global water cycle: Review and synthesis. Journal of Hydrology 319:83–95.

Hu Y, Maskey S, Uhlenbrook S, Zhao H (2011) Streamflow trends and climate linkages in the source region of the Yellow River, China, Hydrol. Process., 25, 3399–3411, doi:10.1002/hyp.8069.

Hu Y, Maskey S, Uhlenbrook S (2012) Trends in temperature and precipitation extremes in the Yellow River source region, China, Climatic Change, 110, 403–429, doi:10.1007/s10584-011-0056-2.

Hu Y, Maskey S, Uhlenbrook S (2013) Downscaling daily precipitation over the Yellow River source region in China: a comparison of three statistical downscaling methods, Theor. Appl. Climatol., 112, 447–460, doi:10.1007/s00704-012-0745-4.

Hu Y, Maskey S, Uhlenbrook S (2013) Expected changes in future temperature extremes and their elevation dependency over the Yellow River source region, Hydrol. Earth Syst. Sci., 17, 2501–2514, doi:10.5194/hess-17-2501-2013.

Immerzeel WW, Van Beek LPH, Bierkens MFP (2010) Climate change will affect the Asian water towers, Science, 328 (5984), 1382–5.

Immerzeel WW, Van Beek LPH, Konz M, Shrestha AB, Bierkens MFP (2012a) Hydrological response to climate change in a glacierized catchment in the Himalayas, Climatic Change, 110, 721-736.

Immerzeel WW, Pellicciotti F, Shrestha, AB (2012b) Glaciers as a Proxy to quantify the spatial distribution of precipitation in the Hunza basin, Mountain Research and Development, 32 (1), 30-38.

IPCC (2000) Special Report on Emission Scenarios, Cambridge Univ. Press, New York.

IPCC (2007) Intergovernmental Panel on Climate Change: Climate change 2007: the Physical Science Basis Summary for Policy makers Contribution of Working Group I to the Fourth Assessment Report of the Intergovernmental Panel on Climate Change, Cambridge University Press, Cambridge, 996 pp..

IPCC (2012) Special Report on Managing the Risks of Extreme Events and Disasters to Advance Climate Change Adaptation (Cambridge: Cambridge University Press).

IPCC (2013) Summary for Policymakers. In: Climate Change 2013: The Physical Science Basis, Contribution of Working Group I to the Fifth Assessment Report of the Intergovenmental Panel on Climate Change. Cambridge University Press, Cambridge,United Kingdom and New York, NY, USA.

Jasper K, Calanca P, Fuhrer J (2006) Changes in summertime soil water patterns in complex terrain due to climatic change, J. Hydrol., 327, 550–563.

Jin H, He R, Cheng G, Wu Q, Wang S, Lü L, Chang X (2009) Changes in frozen ground in the Source Area of the Yellow River on the Qinghai–Tibet Plateau, China, and their eco-environmental impacts. Environ. Res. Lett. 4 (2009) 045206 (11pp)

Katz RW, Brown BG (1992) Extreme events in a changing climate: variability is more important than averages. Clim Change 21:289–302.

Karabörk MC, Kahya E (2009) The links between the categorised Southern Oscillation indicators and climate and hydrologic variables in Turkey. Hydrological Processes 23: 1927–1936.

Kalnay E, Kanamitsu M, Kistler R, Collins W, Deaven D, Gandin L, Iredell M, Saha S, White G, Woollen J, Zhu Y, Leetmaa A, Reynolds B, Chelliah M, Ebisuzaki W, Higgins W, Janowiak J, Mo KC, Ropelewski C, Wang J, Jenne R, Joseph D (1996) The NCEP-NCAR 40-year reanalysis project. Bull Am Meteor Soc 77:437–471.

Kendall MG (1975) Rank correlation measures. Charles Griffin, London.

Kharin VV, Zwiers FW (2005) Estimating extremes in transient climate change simulations, J. Clim., 18, 1156– 1173.

Kharin VV, Zwiers FW, Zhang XB, Hegerl GC (2007) Changes in temperature and precipitation extremes in the IPCC ensemble of global coupled model simulations, J. Clim, 20(8), 1419–1444.

Kioutsioukis I, Melas D, Zanis P (2008) Statistical downscaling of daily precipitation over Greece. Int J Climatol 28:679–691.

Kirshner S (2005a) Modeling of multivariate time series using hidden Markov models, PhD thesis, University of California, Irvine Kirshner S (2005b) The MVNHMM Toolbox. University of California, Irvine, (http://www.stat.purdue.edu/~skirshne/MVNHMM/) Accessed June 2011.

Kjellström E, Bärring L, Jacob D, Jones R, Lenderink G, and Schär C (2007): Modelling daily temperature extremes: recent climate and future changes over Europe, Climatic Change, 81, 249–265, doi:10.1007/s10584-006-9220-5.

Kruger AC (2006) Observed trends in daily precipitation indices in South Africa: 1910–2004. Int J Climatol 26:2275–2285.

Kundzewicz ZW, Mata LJ, Arnell NW, Döll P, Kabat P, Jiménez B, Miller KA, Oki T, Sen Z, Shiklomanov IA. 2007. Freshwater resources and their management. M.L. Parry, O.F. Canziani, J.P. Palutikof, P.J. van der Linden and C.E.Hanson, (Eds.). Climate Change 2007: Impacts, Adaptation and Vulnerability. Contribution of Working Group II to the Fourth Assessment Report of the Intergovernmental Panel on Climate Change, Cambridge University Press, Cambridge, UK: 173–210.

Kundzewicz ZW, Ulbrich U, Brucher T, Graczyk D, Kruger A, Leckebusch GC,Menzel L, Pinskwar I, Radziejewski M, Szwed M (2005) Summer floods in central Europe—climate change track? Nat Hazards 36:165–189.

Kunkel KE, Andsager K, Easterling DR (1999) Long-term trends in extreme precipitation events over coterminous United States and Canada. J Climate 12:2515–527.

Kyselý J (2008) Trends in heavy precipitation in the Czech Republic over 1961–2005. Int J Climatol 29:1745–1758. doi:10.1002/joc.1784.

Lan YC, Zhao GH, Zhang YN, Wen J, Liu JQ, Hu XL (2010) Responses of runoff in the source region of the Yellow River to climate warming. Quaternary international. 226: 60–65. DOI: 10.1016/j.quaint.2010.03.006.

Liang S, Ge S, Wan L, Zhang J (2010) Can climate change cause the Yellow River to dry up? Water Resources Research 46: W02505. DOI: 10.1029/2009WR007971.

Liang SH, Wan L, Zhang JF, Xu DW (2007) Periodic variation in baseflow and its causes in the headwater region of the Yellow River. Progress in Nature Science 17(9): 1222–1228 (in Chinese).

Lindstrom G, Bergstrom S (2004) Runoff trends in Sweden 1807–2002. Hydrological Sciences Journal 49(1): 69–83.

Liu X, Chen B (2000) Climatic warming in the Tibetan plateau during recent decades, International Journal of Climatology, 20, 1729–1742.

Liu X, Yin ZY, Shao X, Qin N (2006) Temporal trends and variability of daily maximum and minimum, extreme temperature events, and growing season length over the eastern and central Tibetan Plateau during 1961–2003, J. Geophys. Res., 111, D19109, doi:10.1029/2005JD006915.

Liu L, Liu Z, Ren X, Fischer T, Xu Y (2011) Hydrological impacts of climate change in the Yellow River Basin for the 21st century using hydrological model and statistical downscaling model, Quaternary International, 244, 211-220.

Liu XY(2004) Runoff variations in the Yellow River source region. Conference on runoff and ecology variations in the Yellow River source region, Zhengzhou, China (in Chinese).

Liu X, Chen B (2000) Climatic warming in the Tibetan plateau during recent decades. International Journal of Climatology 20: 1729–1742.

Liu X, Cheng Z, Yan L, Yin Z (2009) Elevation dependency of recent and future minimum surface air temperature trends in the Tibetan Plateau and its surroundings, Global Planet. Change, 68, 164–174.

Liu Z, Xu Z, Charles SP, Fu G, Liu L (2011) Evaluation of two statistical downscaling models for daily precipitation over an arid basin in China. Int J Climatol. doi:10.1002/joc.2211.

Liu W, Fu G, Liu C, Charles SP (2012) A comparison of three multi-site statistical downscaling models for daily rainfall in the North Chain Plain. Theor Appl Climatol. doi:10.1007/S00704-012-0692-0.

Liu Y, Li X, Zhang Q, Guo YF et al (2010) Simulation of regional temperature and precipitation in the past 50 years and the next 30 years over China. Quatern Int 212:57–63. doi:10.1016/j.quaint.2009.01.007

Liu C, Allan RP (2013) Observed and simulated precipitation responses in wet and dry regions 1850–2100. Environ. Res. Lett. 8: 034002 (11pp).

Lu AM, Jia SF, Yan HY, Yang GL (2009) Temporal variations and trend analysis of the snowmelt runoff timing across the source regions of the Yangtze River, Yellow River and Lancang River. Resources science 31(10): 1704–1709 (in Chinese). DOI: CNKI: SUN: ZRZY.0.2009-10-014.

Lu A, Kang S, Li Z, Theakstone W (2010) Altitude effects of climatic variation on Tibetan Plateau and its vicinities, J. Earth Sci., 21, 189–198.

López-Moreno J, Vicente-Serrano S, Angulo-Martínez M, Beguería S, Kenawy A (2009) Trends in daily precipitation on the northeastern Iberian Peninsula, 1955–2006. Int J Climatol 30:1026–1041. doi:10.1002/joc.1945.

Mann HB (1945) Non-parametric tests against trend. Econometrica 13:245–259.

Manton MJ, Della-Marta PM, Haylock MR, Hennessy KJ, Nicholls N, Chambers LE, Collins DA, Daw G, Finet A, Gunawan D, Inape K, Isobe H, Kestin TS, Lefale P, Leyu CH, Lwin T, Maitrepierre L, Ouprasitwong N, Page CM, Pahalad J, Plummer N, Salinger MJ, Suppiah R, Tran VL, Trewin B, Tibig I, Yee D (2001) Trends in extreme daily rainfall and temperature in Southeast Asia and the South Pacific: 1961–1998. Int J Climatol 21:269–284.

Maraun D, Wetterhall F, Ireson AM, Chandler RE, Kendon EJ, WidmannM, BrienenS, RustHW, SauterT, ThemeßI, Venema VKC, Chun KP, Goodess CM, Jones RG, Onof C, Vrac M, Thiele-Eich I (2010) Precipitation downscaling under climate change: recent development to bridge the gap between dynamical models and the end user. Rev Geophys 48:RG3003. doi:10.1029/2009RG000314.

Marengo JA (2009) Long-term trends and cycles in the hydrometeorology of the Amazon basin since the late 1920s. Hydrological Processes 23: 3236–3244.

Marengo JA, Rusticucci M, Penalba O, Renom M (2010) An intercomparison of observed and simulated extreme rainfall and temperature events during the last half of the twentieth century: part 2: historical trends, Climatic Change, 98, 509–529.

Maskey S, Uhlenbrook S, Ojha S (2011) An analysis of snow cover changes in the Himalayan region using MODIS snow products and in-situ temperature data, Climatic Change, 108, 391-400, doi:10.1007/s10584-011-0181-y.

Masih I, Uhlenbrook S, Maskey S, Smakhtin V (2010) Streamflow trends and climate linkages in the Zagros Mountains, Iran. Clim Change 104:317–338. doi:10.1007/s10584-009-9793-x.

Mehrotra R, Sharma A, Cordery I (2004) Comparison of two approaches for downscaling synoptic atmospheric patterns to multisite precipitation occurrence. J Geophys Res 109:D14107. doi:10.1029/2004JD004823.

Meehl GA, Arblaster JM, Tebaldi C (2005) Understanding future patterns of increased precipitation intensity in climate model simulations. Geophys Res Lett 32:L18719. doi:10.1029/2005GL023680.

Michaels PJ, Knappenberg PC, Fraunfeld OW, Davis RE (2004) Trends in precipitation on the wettest days of the year across contiguous USA. Int J Climatol 24:1873–1882.

Mirza MMQ (2003) Climate change and extreme weather events: can developing countries adapt? Climate Policy 3:233–248.

Milly PCD, Betancourt J, Falkenmark M, Hirsch RM, Kundzewicz ZW, Lettenmaier DP, Stouffer RJ (2008) Stationarity is dead: whither water management? Science 319: 573–574.

Mudelsee M, Börngen M, Tetzlaff G, Grünewald U (2003) No upward trend in the occurrence of extreme floods in central Europe. Nature 425: 166–169.

Moberg A, Jones PD (2005) Trends in indices for extremes in daily temperature and precipitation in central and western Europe, 1901–99. Int J Climatol 25:1149–1171.

Monteith JL (1975) Vegetation and the Atmosphere, Vol. 1: Principles, Academic, London.

Nash JE, Sutcliffe JV (1970) River flow forecasting through conceptual models. Part 1—A discussion of principles, J. Hydrol., 10, 282–290.

Nandintsetseg B, Greene S, Goulden CE (2007) Trends in extreme daily precipitation and temperature near Lake Hövsgöl. Int J Climatol 27: 341–347.

Niu T, Chen LX, Zhou ZJ (2004) The characteristics of climate change over the Tibetan plateau in the last 40 years and the detection of climatic jumps. Adv Atmos Sci 21(2):193–203.

Novotny EV, Stefan HG (2006) Stream flow in Minnesota: indicator of climate change. Journal of Hydrology 334: 319–333.

Osborn TJ, HulmeM, Jones PD, Basnett TA (2000) Observed trends in the daily intensity of United Kingdom precipitation. Int J Climatol 20:347–364.

Polson D, Hegerl GC, Allan RP, Sarojini BB (2013) Have greenhouse gases intensified the contrast between wet and dry regions? Geophys. Res. Lett., 40, 4783-787, doi:10.1002/grl.50923.

Plummer N, Salinger MJ, Nicholls N, Suppiah R, Hennessy KJ, Leighton RM, Trewin B, Page CM, Lough JM (1999) Changes in climate extremes over the Australian region and New Zealand during the twentieth century. Clim Change 42:183–202.

Prudhomme C, Reynard N, Crooks S (2002) Downscaling of global climate models for flood frequency analysis: where are we now? Hydrol Process 16:1137–1150.

Qin J, Yang K, Liang S, Guo X. (2009) The altitudinal dependence of recent rapid warming over the Tibetan Plateau, Climatic Change, 97, 321–327.

Ramos MC, Martinez-Cassanovas JA (2006) Trends in precipitation concentration and extremes in the Mediterranean Penédes-Anoia Region, NE Spain. Clim Change 74:457–

474 Richards, L.: Capillary conduction of liquids through porous mediums, Physics, 1, 318–333, 1931.

Ramanathan V, Crutzen PJ, Kiehl, JT, Rosenfeld D (2001) Atmosphere – Aerosols, climate, and the hydrological cycle Science, 294, 2119–2124, doi:10.1126/science.1064034,.

Rangwala I, Miller J, Xu M (2009) Warming in the Tibetan Plateau: possible influences of the changes in surface water vapor, Geophys. Res. Lett., 36, L06703, doi:10.1029/2009GL037245.

Rangwala I, Miller J (2012) Climate change in mountains: a review of elevation-dependent warming and its possible causes, Climatic Change, 114, 527–547, doi:10.1007/s10584-012-0419-3,.

Rangwala I, Miller J, Russell G, Xu M (2010) Using a global climate model to evaluate the influences of water vapor, snow cover and atmospheric aerosol on warming in the Tibetan Plateau during the twenty-first century, Clim. Dynam., 34, 859–872.

Rössler O, Diekkrüger B, Löffler J (2012) Potential drought stress in a Swiss mountain catchment—Ensemble forecasting of high mountain soil moisture reveals a drastic decrease, despite major uncertainties, Water Resour. Res., 48, W04521, doi:10.1029/2011WR011188.

Salinger MJ, Griffiths GM (2001) Trends in New Zealand daily temperature and rainfall extremes. Int J Climatol 21:1437–1452.

Sato Y, Ma XY, Xu JQ, Matsuoka M, Zheng HX, Liu CM and Fukushima Y (2008) Analysis of long–term water balance in the source area of the Yellow River basin. Hydrol Process 22:1618–1629.

Schaner N, Voisin N, Nijssen B, Lettenmaier DP (2012) The contribution of glacier melt to streamflow, Environ. Res. Lett., 7, 034029(8pp).

Schoof JT, Shin, DW, Cocke S, LaRow TE, Lim YK, O'Brien, JJ (2009) Dynamically and statistically downscaled seasonal temperature and precipitation hindcast ensembles for the southeastern USA, Int. J. Climatol., 29, 243–257, doi:10.1002/joc.1717.

Schmidli J, Frei C (2005) Trends of heavy precipitation and wet and dry spells in Switzerland during the 20th century. Int J Climatol 25:753–771.

Schmidli J, Frei C, Vidale PL (2006) Downscaling from GCM precipitation: a benchmark for dynamic and statistical downscaling methods. Int J Climatol 26:679–689

Schulla J (2012) Model Description WaSiM (Water balance Simulation Model), Tech. rep., Hydrology Software Consulting J. Schulla, Z¨ urich, Switzerland.

Sen PK (1968) Estimates of the regression coefficient based on Kendall's tau. J Am Stat Assoc 63:1379–1389.

Sen Roy S, Balling RC (2004) Trends in extreme daily precipitation indices in India. Int J Climatol 24:457–466.

Shrestha AB, Aryal R (2011) Climate change in Nepal and its impact on Himalayan glaciers, Reg. Environ. Change, 11 (Suppl 1), S65–S77, doi:10.1007/s10113-010-0174-9.

Suppiah R, Hennessy KJ (1998) Trends in total rainfall, heavy rain events and number of dry days in Australia, 1910–1990. Int J Climatol 10:1141–1164.

Sun Y, Solomon S, Dai A, Portmann RW (2006) How often does it rain? J Climate 19:916–934.doi:10.1175/JCLI3672.1.

Tang Q, Oki T, Kanae S, Hu H (2008) A spatial analysis of hydro-climatic and vegetation condition trends in the Yellow River basin. Hydrol Process 22:451–458.

Tebaldi C, Hayhoe K, Arblaster J, Meel G (2006) Going to the extremes: an intercomparison of model-simulated historical and future changes in extreme events, Climatic Change, 79, 185–211, doi:10.1007/s10584-006-9051-4.

Thiel H (1950) A rank-invariant method of linear and polynomial regression analysis, Part 3. In Proceedings of Koninalijke Nederlandse Akademie Van Weinenschatpen A, vol. 53, 1397–1412.

Tripathi S, Srinivas VV, Nanjundiah RS (2006) Downscaling of precipitation for climate change scenarios: a support vector machine approach. J Hydrol. doi:10.1016/j.jhydrol.2006.04.030.

Tryhorn L, DeGaetano A (2010) A comparison of techniques for downscaling extreme precipitation over the Northeastern United States. Int J Climatol. doi:10.1002/joc.2208.

Tu M (2006) Assessment of the effects of climate variability and land use change on the hydrology of the Meuse River Basin. PhD thesis, UNESCO-IHE Institute for Water Education, Delft/Vrije Universiteit, Amsterdam. Taylor & Francis Group plc: London.

Uhlenbrook S (2009) Climate and man-made changes and their impacts on catchments. In Water Policy 2009, Water as Vulnerable and Exhaustible Resources, Proceedings of the Joint Conference of APLU and ICA, 23–26 June 2009, Kovar P, Maca P, Redinova J (eds). Prague: Czech Republic; page 81–87.

Ulbrich U, Brücher T, Fink AH, Leckebusch GC, Krüger A, Pinto JG (2003) The central European floods of August 2002. Part I: rainfall periods and flood development. Weather 58: 371–377.

van Genuchten MT (1980) A closed-form equation for predicting the hydraulic conductivity of unsaturated soils, Soil Sci. Soc. Am. J., 44, 892–898.

Vincent LA, Mekis E (2006) Changes in daily and extreme temperature and precipitation indices for Canada over the twentieth century. Atmos-Ocean 44:177–193

Viviroli D, Archer DR, Buytaert W, Fowler HJ, Greenwood GB, Hamlet AF, Huang Y, Koboltschnig G, Litaor MI, López-Moreno JI, Lorentz S, Schädler B, Schreier H, Schwaiger K, Vuille M, Woods R (2011) Climate change and mountain water resources: overview and recommendations for research, management and policy, Hydrol. Earth Syst. Sci., 15, 471-504, doi:10.5194/hess-15-471-2011.

von Storch H, Navarra A (eds) (1995) Analysis of climate variability. Springer, New York.

von Storch H (1999) On the use of "Inflation" in statistical downscaling. J Climate 12:3505–3506.

Wagner D (1996) Scenarios of extreme temperature events. Clim Change 33:385–407.

Wang G, Qian J, Cheng G, Lai Y (2001) Eco-environmental degradation and causal analysis in the source region of the Yellow River. Environ Geol 40:884–890.

Wang X, Zhou W, Wang D, Wang C (2013) The impact of the summer Asian Jet Stream biases on surface air temperature in mid-eastern China in IPCC AR4 models, Int. J. Climatol., 33, 265–276, doi:10.1002/joc.3419.

Wang W, Chen X, Shi P, van Gelder PHAJM (2008) Detecting changes in extreme precipitation and extreme streamflow in the Dongjiang River Basin in southern China. Hydrol Earth Syst Sci 12:207–221.

Wang G, Qian J, Cheng G (2001) Eco-environmental degradation and causal analysis in the source region of the Yellow River. Environ Geol 40, 884–890.

Wang G (2009) Detection for climate variation trends in the upper Yellow River and its impact on eco-hydrology. Post doctor research report (0701026c). Nanjing Hydraulic Research Institute, China.

Wang Y, Zhou L (2005) Observed trends in extreme precipitation events in China during 1961–2001 and the associated changes in large-scale circulation. Geophys Res Lett 32:L9707. doi:10.1029/2005GL022574.

Wetterhall F, Bárdossy A, Chen D, Halldin S, Xu C (2006) Daily precipitation downscaling techniques in three Chinese regions, Water Resour Res 42:W11423, doi:10.1029/2005WR004573.

Wetterhall F, Halldin S, Xu CY (2007) Seasonality properties of four statistical downscaling methods in central Sweden, Theor. Appl. Climatol., 87, 123–137, doi:10.1007/s00704-005-0223-3.

Wilby RL, Wigley TML (1997) Downscaling general circulation model output: a review of methods and limitations. Progr Phys Geogr 21:530–548.

Wilby RL, Wigley TML (2000) Precipitation predictors for downscaling: observed and general circulation model relationships. Int J Climatol 20(6):641–661.

Wilby RL, Wigley TML, Conway D, Jones PD, Hewitson BC, Main J, Wilks DS (1998) Statistical downscaling of general circulation model output: a comparison of methods. Water Resour Res 34:2995–3008.

Wilby RL, Dawson CW, Barrow EW (2002) SDSM—a decision support tool for the assessment of regional climate change impacts, Environ Modell Softw 17(2):145–157.

Wilby RL, Dawson CW (2013) The statistical DownScaling Model: insights from one decade of application, Int. J. Climatol., 33, 1707–1719, doi:10.1002/joc.3544.

Wilby RL, Tomlinson OJ, Dawson CW (2003) Multi-site simulation of precipitation by conditional resampling. Clim Res 23:183–194.

Wu P, Christidis N, Stott P (2013) Anthropogenic impact on Earth's hydrological cycle. Nature Climate Change 3: 807–810.

Xie C, Ding Y, Liu S (2004) Changes of climate and hydrologic environment for the last 50 years in the source region of Yangtze and Yellow River. Ecology and Environment, 13(4):520-523.

Xu C (1999) Climate change and hydrological models: a review of existing gaps and recent research developments. Water Resour Manag 13(5):369–382.

Xu Y, Xu C, Gao X, Luo Y (2009a) Projected changes in temperature and precipitation extremes over the Yangtze River Basin of China in the 21st century. Quaternary Int 208:44–52.

Xu Z, Li J, Liu C (2007) Long-term trend analysis for major climate variables in the Yellow River Basin. Hydrol Process 21:1935–1948.

Xu Z, Zhao F, Li J (2009b) Response of streamflow to climate change in the headwater catchment of the Yellow River basin, Quatern. Int., 208, 62–75.

Xu C, Luo Y, Xu Y (2011) Projected changes of precipitation extremes in river basins over China. Quaternary Int 244:149–158.

Yang Z (1991) Glacier Water Resources in China (Lanzhou: Gansu Sci. and Technol.) p 158.

Yang D, Li C, Hu H, Lei Z, Yang S, Kusuda T, Koike T, Musiake K (2004) Analysis of water resources variability in the Yellow River of China during the last half century using historic data. Water Resour Res 40:W06502. doi:10.1029/2003WR002763.

Yang JP, Ding YJ, Liu SY, Liu JF (2007) Variations of snow cover in the source region of the Yangtze and Yellow River in China between 1960 and 1999. Journal of Glaciology 53(182): 420–426.

Yang C, Chandler RE, Isham VS, Wheater HS (2005) Spatial-temporal rainfall simulation using generalised linear models. Water Resour Res 41:W11415.

Yang JS, Chung ES, Kim SU, Kim TW (2012) Prioritization of water management under climate change and urbanization using multi-criteria decision making methods, Hydrol. Earth Syst. Sci., 16, 801–814, doi:10.5194/hess-16-801-2012.

Yang T, Hao X, Shao Q, Xu C, Zhao C, Chen X, Wang W (2012) Multi-model ensemble projections in temperature and precipitation extremes of the Tibetan Plateau in the 21st century. Global Planet Change 80–81:1–13.

Ye B, Li C, Yang D, Ding Y, Shen Y (2005) Variation trend of precipitation and its impact on water resources in China during last 50 years (II): Monthly variation. Journal of Glaciology and Geocryology 27(1): 100–105.

You Q, Kang S, Pepin N, Yan Y(2008) Relationship between trends in temperature extremes and elevation in the eastern and central Tibetan Plateau, 1961–2005, Geophys. Res. Lett., 35, L04704, doi:10.1029/2007GL032669.

Young GJ, Hewitt K (1990) Hydrology research in the upper Indus basin, Karakoram Himalaya, Pakistan. IAHS Publications, 190, 139–152.

Yu R, Wang B, Zhou T (2004) Tropospheric cooling and summer monsoon weakening trend over East Asia. Geophys Res Lett 31:L22212. doi:10.1029/2004GL021270.

Zhai P, Pan X (2003) Trends in temperature extremes during 1951–1999 in China. Geophys Res Lett 30(17):1913. doi:10.1029/2003GL018004.

Zhai P, Sun A, Ren F, Liu X, Gao B, Zhang Q (1999) Changes of climate extremes in China. Clim Change 42:203–218.

Zhai P, Zhang X, Wan H, and Pan X (2005) Trends in total precipitation and frequency of daily precipitation extremes over China. J Climate 18(6):1096–1107.

Zhai J, Zeng X, Su B (2009) Patterns of dryness/wetness in China before 2050 projected by the ECHAM5 model. Adv Clim Change Res 5:220–225.

Zhang Y, Kuang X, Guo W, Zhou T (2006) Seasonal evolution of the upper-tropospheric westerly jet core over East Asia. Geophys Res Lett 33:L11708. doi:10.1029/2006GL026377.

Zhang X, Hogg WD, Mekis E (2001) Spatial and temporal characteristics of heavy precipitation events over Canada. J Climate 14:1923–1936.

Zhang X, Zwiers FW, Li G (2004) Monte Carlo experiments on the detection of trends in extreme values. J Climate 17:1945–1952.

Zhang Q, Xu CY, Zhang Z, Ren G, Chen YD (2008a) Climate change or variability? The case of Yellow river as indicated by extreme maximum and minimum air temperature during 1960–2004. Theor Appl Climatol 93:35–43.

Zhang X, Raghavan S, Bekele D, Hao F (2008b). Runoff Simulation of the Headwaters of the Yellow River Using the SWAT Model With Three Snowmelt Algorithms. Journal of the American Water Resources Association (JAWRA) 44(1):48-61. DOI: 10.1111 / j.1752-1688.2007.00137.x.

Zhang X, Harvey D, Hogg WD, Yuzyk TH (2001) Trends in Canadian streamflow. Water Resources Research 37: 987–998.

Zhang Q, Jiang T, Germmer M, Becker S (2005) Precipitation, temperature and discharge analysis from 1951 to 2002 in the Yangtze River basin, China. Hydrological Sciences Journal 50(1): 65–80.

Zhao FF, Xu ZX and Huang JX (2007) Long-term trend and abrupt change for major climate variables in the upper Yellow River Basin. Acta Meteorol Sin 21(2):204–214.

Zheng HX, Zhang L, Liu CM, Shao QX, Fukushima Y (2007) Changes in stream flow regime in headwater catchments of the Yellow River basin since the 1950s. Hydrol Process 21:886–893.

Zhou TJ, Yu RC (2006) Twentieth-century surface air temperature over China and the globe simulated by coupled climate models, J. Climate, 19, 5843–5858, doi:10.1175/JCLI3952.1.

About the author

Yurong HU was born in 1974 in Henan Province, China. She obtained her BSc degree in hydrology and water resources in 1996 from Hohai University, China. From October 2005 to April 2007 she studied MSc in hydrology and water resources in UNESCO-IHE Institute for Water Education, the Netherlands. She has extensive experience in hydrology, water resources and river basin management. Since June 1996 she has been working as a water resources manager at Yellow River Conservancy Commission (YRCC), China, and she is responsible for a number of projects regarding water policy, water price, water license, water allocation and regulation, water resources planning and water use assessment in the Yellow River. In 2009, she started a (part time) PhD at UNESCO-IHE Institute for Water Education, Delft, the Netherlands. Her PhD research focused on the impacts of climate change on hydrology and water resources in the Yellow River source region, China.

Her key areas of expertise and interest are integrated water resources management, statistical hydrology, process-based distributed hydrological modeling, statistical downscaling, and climate change impacts and adaptation at a river basin scale. She has co-authored a number of scientifically important and practically relevant papers and has presented her research at various national and international workshops and conferences. Yurong HU is a frequent reviewer of manuscripts submitted to international scientific journals, including, for example, International Journal of Climatology and Hydrology and Earth System Science.

She is currently working for YRCC as a senior engineer in water resources management. She is married to Lunshun Wang and they have a son Dinghan Wang.

Selected publications

Hu Y, Maskey S, Uhlenbrook S, Zhao H (2011) Streamflow trends and climate linkages in the source region of the Yellow River, China, Hydrol. Process., 25, 3399–3411, doi:10.1002/hyp.8069.

Hu Y, Maskey S, Uhlenbrook S (2012) Trends in temperature and precipitation extremes in the Yellow River source region, China, Climatic Change, 110, 403–429, doi:10.1007/s10584-011-0056-2.

Hu Y, Maskey S, Uhlenbrook S (2013) Downscaling daily precipitation over the Yellow River source region in China: a comparison of three statistical downscaling methods, Theor. Appl. Climatol., 112, 447–460, doi:10.1007/s00704-012-0745-4.

Hu Y, Maskey S, Uhlenbrook S (2013) Expected changes in future temperature extremes and their elevation dependency over the Yellow River source region, Hydrol. Earth Syst. Sci., 17, 2501–2514, doi:10.5194/hess-17-2501-2013.

Hu Y, Maskey S, Uhlenbrook S (2014) Impacts of climate change on the hydrology of the Yellow River source region, China, Submitted to Climatic Change.

Hu Y, Chen Y, Li Y (2003), Enhance Yellow River water management with water right theory, Proceedings of the first international Yellow River Forum, Zhengzhou, China.

Hu Y (2004) Challenges encountered by unified management and control of water resources in the Yellow River and its countermeasures, China Water Resources 1 (in Chinese).

Hu Y, Chen Y (2004) Theory and practice of water right transfer in the Yellow River, China Water Resources 15 (in Chinese).

Chen Y, Su Q, Hu Y (2007), Promoting water saving society establishment through water right transfer, China Water Resources 19 (in Chinese).

Hu Y, Wang J, Li Y (2007), Promoting integrated water resources management in the Yellow River with lessons from the Netherlands, Proceedings of the third international Yellow River Forum, Zhengzhou, China.

Zhang W, Hu Y, Chen L (2007), The study on the duration of water right transfer in the Yellow River, China Water Resources 19 (in Chinese).

T - #1026 - 101024 - C134 - 240/170/7 - PB - 9781138027145 - Gloss Lamination